There Might be HOPE

M. E. Marler

Copyright 2023 @ M. E. Marler

All rights reserved. No portion of this publication may be reproduced, stored in a retrieval system, or transmitted in any form or by any means electronic, mechanical, photocopying, or otherwise without the prior pemission in writing of the author. The scanning, uploading, and distribution of this book via the Internet or via any other means without the written permission of the author is illegal and punishable by law.

ISBN: 9798363630705

DEDICATION

This book is dedicated to the glory of God, my Father,

Jesus Christ, my Savior, and my dad.

Contents

Preface i

Acknowledgements vi

Chapter 1 Dust to Dust 1

Chapter 2 Faith Until Sight 19

Chapter 3 Native Farmers 39

Chapter 4 Fast Kill, Slow Kill 55

Chapter 5 An Unloved Flower 67

Chapter 6 The Diversified Farmer 77

Chapter 7 Dead Zones 91

Chapter 8 Apex Omnivore 101

Chapter 9 The Forests of Arabia 119

Chapter 10 All Things New 141

Notes 153

Preface

I have never aspired to write a book and am always amazed at the number of people who've expressed to me their own eagerness to do so. My dad actually began this book many years ago. It was his third book, and his first and only nonfiction work. None have been published. An entrepreneur, visionary, voracious reader, and lifelong learner, Dad had co-founded an organic fertilizer company in the Pacific Northwest as I was studying agriculture at Texas A&M. Like many entrepreneurs, Dad's ultimate dream was for his kids to come work with him. I had been teaching agricultural science in an urban Texas high school for a couple of years when Dad's health and mainly his heart began to fail. Initially, Dad recruited me to help him set up educational workshops, something I had previous experience and expertise in. The workshops would share what he had learned about the soil microbiome with the greater agricultural community. It later came out that Dad...happened to be writing a book he wanted to sell at his workshops. He just needed me to edit it, and then, I could get started setting up workshops. Educational workshops, I knew. Professional editing, I did not. The next thing I knew, I was sitting on the floor of Barnes and Nobles combing through "The Writer's Market," and discovering that the literary world was far more obsessed with dark, teenage, fantasy fiction pieces than anything to do with soil science.

Had I read Dad's book fresh out of college, I doubt I would have bought into it. During my college years, when I approached professors about organic farming, they generally shrugged their shoulders. One replied to me with, "Well, you don't want to bite into an apple with a worm in it, do you?" It was dismissed as a fad. The speed with which new technology was almost unquestionably accepted ought to have seemed strange and unscientific to me then, as it does to me now, in retrospect. At the time, I simply deferred to my professor's decades of scholarship and expertise and accepted that my father's company was supplying a niche market.

Post college, I went to work as an ag science teacher at a large, urban high school. Along with my regular classes, our school also offered advanced students the opportunity to receive credit for independent study projects. A young lady, with an interest in organic foods, approached me and asked if I would sponsor and supervise a project she had in mind. She wanted to compare organic seeds and fertilizers with conventional. At the time, I already had one student competing around the state in FFA Agri-Science Fairs. Why not have two? She agreed to join our FFA program, and I agreed to supervise her project. I can't remember what, if any, hypothesis or predictions I had at the beginning of her project, but I certainly wasn't expecting the ultimate outcome.

I furnished her with just about every type of vegetable seed that an ag teacher could buy with a Home Depot P.O. along with 2 small bags of conventional and organic fertilizers, also procured from Home Depot. She set up in our partly functional school greenhouse and divided her experiment into four test groups, organic seed with organic fertilizer, organic seed with conventional fertilizer, conventional seed with organic fertilizer, and conventional seed with conventional fertilizer. Everything went well with little observable difference between the groups until Spring Break.

As we neared Spring Break, I repeatedly reminded my student that she wouldn't have access to the greenhouse over the next week. It was especially important that she remembered to water her young seedlings well before she left on Friday.

She didn't. When I went to check on her plants at the end of the day, I made, what in hindsight, was perhaps a fateful decision. I decided not to intervene and allow her to reap the consequences of her choices. I locked up the greenhouse and went home.

Mid-week, I had to return to the school to gather students and materials for a judging competition. Feeling that I had been a bit harsh, my conscience got the better of me, and I went out to check on her plants. Trays of poor, drought-stressed seedlings were crying out for water. I caved. Even as I gently soaked them, two things were abundantly clear. First, regardless of seed type, the organically fertilized seedlings were going to be just fine. Secondly, the conventionally fertilized seedlings were goners. We hadn't used my dad's fertilizer in the experiment, but he was clearly on to something. If organic fertilizers could protect plants against drought stress, when conventional fertilizers couldn't, the implications would be huge. I was a believer.

By the time Monday rolled around, traces of white fungi had emerged on the surface of the organically fertilized soils. Today, I know this fungus as Mycorrhizae, a hero to drought stressed plants. My student refused to take any part in the excitement of this scientific epiphany. In true teenage fashion, she poured forth her woes to just about any listening ear within my earshot, of how her ag teacher tried to drown her plants, her project, and her dreams and actually did kill some of them. The Drought of Spring Break may have been forgotten, but the Deluge ne'er will be. No good deed goes unpunished. Ag teachers will understand.

Candidly, Dad and I had numerous clashes about the book. I suppose that is as normal for an author and an editor as for a father and a daughter. While Dad's brilliant and often prophetic mind had produced a partial manuscript, that I believed was critical for a wide-audience to read, citing his sources wasn't Dad's strong suite. An avid reader, Dad regularly plowed through probably 3-4 books a week, in addition to news and scientific articles. He was enthralled with learning, synthesizing, and sharing what he was learning. Dad was into dreaming, exploring, building, and creating. Dad was not into note taking. By the time I began editing, the scientific articles, books, and journals that had been foundational to Dad's ideas and writing, were virtually impossible to recompile in retrospect.

Despite this, after many months of querying every potential publisher and agent I could find and being bluntly rejected by most, a midwestern agricultural university press requested to see a copy of Dad's manuscript. Delighted, I called Dad to tell him and received the last response I expected from him. Dad suddenly got cold feet. He absolutely refused to let me send it.

We moved on to other projects, and Dad went home to be with the Lord about 5 years later. My conscious wouldn't let the book rest though. My conscious kept convicting me that the knowledge I had been graciously entrusted with, needed to be offered to a wider audience. I knew the book offered life-changing information that could help people understand what was happening with their food, their health, and the environment, and potentially chart a different course than the one we headed down. I picked up where Dad and I had left off.

In my mind, I initially thought of it like finishing out a house. I would paint a room or two, install some appliances, and maybe lay some carpet. Voila! However, as I began the work of bringing the book up to date, I found co-authoring a book with someone who isn't there to be more challenging than I had presumed. What's more, Dad and I have markedly different writing styles. How was I going to merge the two? It became an impossible task. My tools quickly evolved from a paint brush to a sledgehammer. In the end, I knew I had to take the house down to the foundations. It was hard.

So, while the book you are about to read retains many of my father's ideas and some of his flow of thought, the words, and many of ideas, are now entirely my own. His book really became a springboard for this one, and I know he is smiling down from heaven on it.

Like many of my dad's ideas, when he began his book, his ideas were considered fringe, at best. The very idea of a soil microbiome was, not too long ago, were considered laughable and verboden. Like every other idea my dad had ever latched onto, with time, research had increasingly substantiated what Dad instinctively had known. It's important to note that although he was certainly one of the earliest advocates of them, Dad's ideas about the soil microbiome didn't originate with him either. As discussed in the book, a handful of passionate, determined, and brilliant soil scientists toiled and persevered over many decades to find proof of their own hypothesis. Thankfully, an increasingly broad spectrum of scientists' research continues to expand our understanding of the soil microbiome today and validates the perseverance of those who came before. Nevertheless, a substantial disconnect between the groundswell of recent research, public understanding, and mainstream agriculture remains.

My prayer is that God would use this book, like a lighthouse, to alert readers of the perilous coastline that now lies before us and to help them navigate the increasingly treacherous waters we have found ourselves within. Moreover, I hope this book will bless those who read it. I pray that as you read it, you will increasingly see the goodness of God poured out in His creation and upon His people and that ultimately, you would trust in Him who alone has the power to save, to restore, and to make new.

ACKNOWLEDGEMENTS

I would like to thank my mom for her steadfast love, support, presence, and prayers. I also owe my deepest appreciation to everyone who has ever prayed for me. Your prayers made this book possible. May the Lord return the kindness you have shown me 10,000-fold.

Chapter 1 Dust to Dust

"Cursed is the ground for your sake; In toil you shall eat of it all the days of your life.... for out of it you were taken. For dust you are, and to dust you shall return."

~ Genesis 3:17, 19

The Mediterranean glistens as local fishing vessels weighed down with their fresh, morning catches coast in towards land on translucent waves of cobalt and turquoise. Their destination is a magnificent, nearly circular harbor guarded by a towering, 11 story, limestone lighthouse.[1] Massive, quarried stone docks line the harbor perimeter, marrying the sea with a stately, columned portico. Adjoining storehouses overflow with sacks of wheat and sealed, earthen amphora heavy with olive oil and wine. Provincial slaves labor shoulder-to-shoulder with African free-men loading heavier, sea-worthy vessels capable of freighting between 100 and 600 tons of cargo. Powered by simple, cotton sails and the sweat of oarsmen, these ships annually bear some 3 million pounds of olive oil to Rome as tribute in this, the year of 210A.D. Likely, this tribute was a mere sampling of the total olive oil and assorted agricultural commodities produced by the region and headed to locations throughout the Empire[2].

Beyond her cornicopic wharves and warehouses, is the Cardo, a central, life-giving, north-south thoroughfare. This broad, limestone street is bordered by colonnades of ornately sculpted, marble reliefs depicting both mythological figures and hometown heros. Detailed and vibrantly colored tile mosaics chime in boasting of the sea's abundance and the land's bounteous, cultivated crops. To the west of our harbor lie the standard Roman seaside attractions, a magnificent three-storied amphitheater overlooking the Mediterranean, a 16,000-seat gladiator arena, and a chariot racecourse. Nestled amongst them are a plethora of Greek and Roman Temples, hedged by multitudes of luxurious villas. Its exact population is debated, but now, at her zenith, the city is home to at least 30,000[3].

At the city's heart lies a flourishing marketplace fashioned from blue and golden-veined marble columns. Here regional commodities of olive oils, wines, and wheat change hands alongside pastured beef, wild-caught and farmed fishes, wild boar, venison, and an array of poultry including ducks and geese fattened for foi gras. Fresh, local, and inherently organic produce accompany these staples by season, and boast dozens of varieties of pears, apples, peaches, nectarines, sweet cherries, watermelon, pomegranates, peas, turnips, and kale along with onions, beans, and lentils[4]. Meat may be a luxury for the wealthy, but the sheer abundance of fruits and vegetables enables even the poorest to afford their fill of them.

To the city's West and Northwest, lie its literal bread and butter, its farmland. Long before Roman rule, farming flourished in this area, but under it, agriculture intensified and boomed. Forests and grasslands whose roots stretched down into rich, dark soils once dominated the landscape here and were slowly broken up by intensive agriculture beginning around 1 B.C. Now in 210 A.D., farms blanket the countryside. Olive groves dominate the hillsides while valleys cradle grain, vegetable crops, and grazing pastures for cattle. Modern agriculturists and researchers debate how much a single olive tree produced in Roman times but agree that a number of between 66 and 220 lbs. of olive oil per tree is likely. Pliny, though, called these trees "millennia" expecting them to annually produce around 720 lbs. (1,000 libra) of olives each year[5], a figure comparable with modern high-yielding producers, notably without modern methods.

 Despite their immense yearly tribute in olive oil, according to Pliny, the area was best known for its cereal production. Wheat production made our city, Leptis Magna, and many similar Roman agricultural colonies economic powerhouses and fueled Rome and her endless military campaigns. Military strategy, prowess, and personality frequently steal the spotlight from the quiet foundation undergirding every successful campaign, a reliable food supply. Apart from the collective agricultural strength of colonies like Leptis, the Roman Empire could not and would not have existed.

 The Romans begged, borrowed, and stole much of their agricultural acumen from preceding and surrounding cultures. Legume cover crops, manure fertilizers, crop rotations, fallowing fields, and diversified farming complimented technology such as remarkable iron-wheeled plows, irrigation systems, ox-drawn harvesters, and even greenhouses. This collective knowledge and practice outproduced both previous, known, farming generations as well as those to come for millennia. Vineyards, grain fields, and olive groves produced robust yields that would not begin to be met nor rivaled until the Green Revolution of the 1950's nearly 2,000 years later.

By way of the Mississippi River, it's 102 miles from the Gulf of Mexico to the Port of New Orleans. Broad, blue skies frame the glassy, shimmering, olive green ship channel that begins at the Mississippi Delta's Southwest Pass. For nearly 200 hundred years, ship captains have depended upon a successive series of lighthouses to guide them and their valuable cargo into the Pass. The eldest of these is an 1839 vintage, white-washed brick tower that despite gradually sinking since its inception, still stands today. Within its sight watches its successor. This 13-story, cast iron, concrete, and timber fortress was a cutting-edge feat of late 19th century engineering. The third and final manned lighthouse is no more. Constructed in the 1960's with all the bells and whistles of a contemporary offshore oil rig including living quarters and a helicopter pad, it fell to Hurricane Katrina's brutality in 2005.

Today, an unmanned, skeletal tower minimally adorned with solar panels and an electronic, flashing light alerts ships to the Pass entrance. Though barely taller than the lighthouse at Leptis Magna, it beacons immense, welded-steel ships freighting between 60,000-100,000 tons of cargo. These behemoths form a nearly continuous, stern to bow, parade up the Mississippi delivering in refrigerated "reefers" vegetables from Mexico, Canada, and China. Fruits and coffees arrive from Costa Rica, Guatemala, and Chile. Bulk cargo ships laden with rice and soybeans hail from Thailand, India, Argentina, Bolivia, and Peru. Inbound to New Orleans they pass, often less than a ship's width apart from their predecessors which having been unloaded and refilled are now bound for China, Mexico, Columbia, Egypt, and Germany carrying rice, soybeans, and corn.

Navigating upriver from the Pass to the Port of New Orleans requires the better part of a day's journey and the onboarding of a skilled, Louisiana river pilot with keen understanding of the ever rolling and shifting sands below the river's smooth surface to keep the ships from running aground. With the careful maneuvering of a tug,

There Might Be Hope

ships parallel park alongside the New Orleans dock, and Microsoft blue gantry cranes roll down rails to position themselves for unloading. Perpendicular to the ship, gantry crane operators hover within tiny, glass pods, some 13 stories up, and mechanically hoist mustard yellow, rusty red, burnt orange, navy, and grey containers from the merchant vessels onto awaiting trucks and railcars bound for distribution centers and ultimately retailers further inland.

Just to the North and East of the port, concrete, steel, and glass skyscrapers reaching 50 floors tall watch over the Port of New Orleans and are hedged by welcoming neighborhoods of red brick, cast iron, and vibrantly painted historic homes. Trolleys sew commercial and residential districts together with riverside attractions like the Mercedes Benz Superdome, St. Louis Cathedral, and Audubon Park. Throughout the city, come the warm, beckoning aromas from rich, dark, simmering rouxs. Fresh bell peppers, onions, and celery are slowly stewed with tender, Gulf crabs and oysters and thickened with flour and butter to create hearty jumbalayas and spicy gumbos. Fresh, cornbread battered shrimp are fried, wrapped in warm, French bread and soaked in Remoulade sauce. Pillowy Beignets are showered in sweet, powered sugar and accompanied with hot, rich coffee. These foods are the life blood of New Orleans nearly 400,000 citizens. The Amazon-acquired chain, Whole Foods, helps to supply the ingredients. Perhaps the most widely recognizable of New Orleans groceries, Whole Foods originally made its name by purchasing and showcasing fresh, local and organic produce. Farmers' stories and portraits once were stylishly featured alongside the fruits of their efforts, but statistically, today, most of America's food is neither fresh nor local. In spite of the local foods movement, U.S. vegetable and grain imports have more than doubled since 1999[6]. In fact, today's typical American meal features "ingredients from at least 5 countries outside of the U.S."[7] Most of America's "fresh" fruits and vegetables have waited anywhere from a few weeks to a year in storage by the time they appear in the supermarket. In the fresh food department, America is more akin to Rome than Leptis Magna, but America still dominates food exportation and particularly, grain exportation. In 2021, almost 60% of the United States grain exports left American soil from the Louisiana port region in reloaded cargo ships bound primarily for China and Mexico, but also Europe, South America, and most certainly, Africa.

Finding modern day Leptis Magna, or what remains, means a roughly 2-hour drive southeast of Tripoli, Libya. Eroded hillsides, pockets of bullet-ridden, rural villages, and vestiges of a cruder oil industry repetitiously narrate the now seemingly endless centuries of economic desolation and political strife Libyans have endured since the fall of Leptis Magna. The mummified remains of our once great city remain mostly buried under shifting sand dunes. The excavated portions though stun and shine in both beauty and complexity of extraordinarily well-preserved stone structures and mosaic artwork. Marble blocks and reclining columns have been strewn about over time from the Grecian style buildings they once conjoined. The market and the harbor's former glorious and immense stocks of grains, olive oil, wine, and wide varieties of locally grown produce are no more. Libyans now import up to 75% of their food,[8] and 95% of Libya today is desert. The former abundance of Leptis Magna's market must be as difficult for modern Libyans to envision as her former residents of its current desolation.

Standing near the crumbling lighthouse foundation as waves pound against its limestone and granite remnants, you may watch an occasional cargo ship or oil tanker powering across the Mediterranean, but today, they dock at other ports. Leptis Magna's harbor no longer brims with sea-worthy ships bearing agricultural commodities. Today, it brims with silt. So much silt fills the harbor that you can often walk completely across it without getting your shoes wet, and until recently, many archaeologists questioned whether this premier harbor ever functioned. Long-standing theory asserted that the Roman empire's decline in military power, along with a terrific earthquake in 365 A.D., wrought this city's ruin and virtual desertion, a theory both comfortable and acceptable to us living in a technological society. Is it true though? Would the same theory carry equal viability for a prosperous agrarian economy where fertile, profitable lands existed to rebuild wealth?

Archaeologists from 2 independent teams have reexamined these ideas and come to some fresh conclusions. Weathered, olive-press remnants, dating back to the Roman era, stand like tombstones amidst sandy dunes, and mark nearly 50 farm and home sites that once flanked the countryside around Leptis Magna at its peak in 200 A.D. Evidence testifies though that by the mid-300's[9], decades prior to the 365A.D. earthquake, nearly half of these settlements had already been abandoned. Like agricultural production across the Roman empire, Leptis Magna's agricultural boom lasted just around 200 hundred years from 100 A.D. to 300 A.D. Despite political pressures, why would fertile and productive lands be abandoned in an agrarian economy? How did North Africa become the barren wasteland and desert we see today? Our answer may likely reside in her harbor and a dammed Wadi.

South and slightly east of Leptis Magna lies the Wadi Lebda. A seasonal, fresh-water river basin that Romans dammed around 70 A.D., the dam stabilized intermittent flows, protected the city from frequent flashfloods, and supplied running water to her citizens, their homes, businesses, and public fountains. Leptis Magna's harbor served as the natural outlet of the Wadi Lebda and married with a second Wadi flowing via aqueducts from Western farmlands. The Wadi Ledba dam stands today as it did then at 6 meters tall and 200 meters long. Romans built a secondary, 5-meter high, exterior barrier wall as a little extra insurance to shield the city should the primary dam give way, but archaeological examination found no evidence of earthquake damage to either structure. In fact, by the 365 earthquake, the dam that once had served as the city's main water supply was itself clearly no longer functional. It was already filled to the brim with silt. Rather than contain water, water would have cascaded over the dam like a waterfall and poured directly into the channel leading to the harbor. Successive layers of the harbor's soil also confirm that the silt was brought in sequentially via the Wadi. Where did the soils come from? They came from the once fertile farms surrounding Leptis Magna. The city wasn't thrown into sudden chaos and destruction by an earthquake. Slowly at first and then rapidly, the soils that built these farms, that powered the harbor, and that fed an empire quietly failed and washed down the Wadis and out into the Mediterranean.

Natural disasters or attacks wrought by warring nomadic tribes are frequently blamed for the mysterious abandonment of these ancient cities. Isn't it more likely that these disasters and attacks were merely the final death blow on a civilization already slowly dying from diminishing soil fertility and weakened from malnutrition? Neither glamorous nor historically well-understood, soil loss is usually a gradual process. Fields usually fail first in part and then one by one across a farm. As topsoils quietly fail and slip away, one by one, farmers are forced to leave their lands as an area plays out. By around 320 A.D., Leptis Magna's once fertile farmland soils had washed down into the wadi and out to sea. Slowly and softly, desert sands blew across and began to bury Leptis Magna leaving the city for future generations to unearth. By the Vandal invasion of 439 A.D., little likely remained capable of sustaining either defenders or marauders. By 600 A.D., only 2 identifiable, outlying farms remained. Thereafter, the area was frequented mostly by nomadic Arabic tribes and later inhabited by subsistence farmers.

Leptis Magna isn't the only Roman harbor adjacent to a once intensive farming community that today brims with silt, and it isn't the lone example of soil depletion preceding a culture's economic doom either. Iraq, Syria, Jordan, Israel, and Lebanon were all once historically high-yielding agrarian and even densely forested areas famous for their oaks, cedars, vineyards, wheat fields, and well-nourished herds of cattle and horses. Today, in many areas they are overseen by washed-out hills and plateaus and lie *"buried up to 13 feet deep under erosional debris*[1011]. If capable of sustaining any agrarian life at all, it is generally limited to sheep and goats, the hardiest and least nutritionally selective of all herd livestock. Long ago, these civilizations ruled the known world until they could no longer feed their inhabitants. Today, these lands are known as deserts, but substantial evidence indicates that they are less desert than desertified.

Most Libyans today live huddled on the small, remaining sliver of coastline that has not been claimed by the ever-expanding Saharan desert. Romantic images of nomadic, Tuareg tribes aboard camels rhythmically striding across rusty sand dunes have become synonymous with the Sahara, but up until 4,000 B.C., the Sahara was a tropical oasis filled with massive lakes and rivers. Even as recently as 2,000 years ago, much of the Sahara was still prime cattle country. How the Saharan desert was born is up for speculation, but soil samples indicate it happened in less than 200 years. What was once a relatively limited desert has expanded exponentially to encompass 8% of our earth's total land mass, and it's still growing.

Leptis Magna may not be in the export business anymore, but the Sahara certainly is. About 400-600 million tons of its most easily mobilized remaining asset depart the Sahara annually. Soil. The majority of the Sahara—80%---never made yesteryear's enchanting post card or today's Instagram feed. Those steeply sloping, iron-rich, dunes tall enough to bury a New Orleans high-rise with sand to spare only cover about 20% of the Sahara today. After 2 millenia of completely exposed soils and severe and persistent winds, most of the Sahara doesn't look like the Sahara anymore. It looks like rocks. Wind-chiseled rock formations, rock piles covering rock mountains, and barren bedrock are the reality of up to 80% of our modern Sahara. Most of the Sahara isn't in the Sahara anymore. Most of it is at the bottom of the Atlantic. A good portion of the remaining sands though continue to reach South and North America, the Caribbean, Europe, the Middle East, and sub-Saharan Africa.

Wind might seem like an equal-opportunity force, but it isn't. Simple wind physics dictates that the heavier you are, the less likely you are to be carried off by it. Farmland topsoils are no different, and unfortunately, our best soils tend to be our lightweights. Silt loams are among the first to get picked up and spirited away. While coarser and heavier sand granules stay behind, these finer soils, along with organic matter and nutrients, are easily snatched up thousands of feet into the atmosphere and carried for thousands of miles more to be deposited hopefully onto other continents, but more frequently into the ocean. Coarser, heavier sands either stay behind or generally travel respectively shorter distances, bouncing along the ground. The result is that wind erosion causes even once prime farmland, over time, to become coarser and coarser, sandier and sandier as farms inch closer and closer to bedrock.

Under the most idyllic climate conditions, forming a single inch of new soil takes 30-200 years. In semi-arid or desert areas such as Leptis Magna, one inch of topsoil formation may take thousands of years or even fail completely. Farming, since its inception, seems to have always inherently produced some level of erosion. Because soil formation is so slow, even low erosion rates are dangerous and unsustainable over the long haul. Losing any more than 2 tons of topsoil per acre per year is considered an irrecoverable loss. Lose a steady 5 tons per acre per year, and you'll have lost 6 inches of topsoil per acre in about 200 years[12]. Most topsoil is only 6 inches deep. That also happens to be the current average erosion rate for farms across the U.S.[13] If we were just beginning to farm in the U.S., this might not be so concerning, but Americans aren't newbie farmers. We're veterans on the back end of that 200 years of loss, and from the 1950's until recently, annual erosion has been well more than twice the current rate. In many parts of the U.S. losing 10 tons per acre per year of farmland or more is still the norm. Conservatively, you might deduce that we have 60 years of farmland topsoil left in the U.S., but realistically, we have 30. That's it. Thirty years of U.S. farmland soils remain. This is an average. More intensively farmed lands such as the Plains states, from North Dakota to Texas, will likely be the first to hit bedrock, followed by the West Coast. Once initiated, erosion typically begets more erosion, and if unchecked, erosion begets desertification. Desertification begets more desertification.

To say that we're losing an average of 5 tons per acre per year of topsoil on farms across the U.S. is an estimate, and honestly, a conservative one. The reason? Measuring wind erosion with 100% accuracy is virtually impossible, and we generally err on the low side. Imagine if you live a half a mile away from your nearest neighbor. After a rainstorm, you compare notes and learn that while the rain gauge at your house measured 1 inch of rain, your neighbor received 3 inches. It seems logical to estimate that the crops midway between your two houses received 2 inches of rain, but unless you also had a rain gauge there you wouldn't know for sure. Maybe the most intense part of the storm passed between your houses and that area received 3.5 inches of rain. Maybe it only received half an inch of rain.

Measuring erosion is similar. The USDA measures soils leaving farms annually through small, sample plots of farmland under varying conditions, but inevitably, the total amount of eroded farmland soils in our rivers, like the Mississippi, is considerably greater than our predictions, based on these test plots, say it ought to be.

It has been said that to understand the enormity of this problem one need only stand at the confluence of the Mississippi and Ohio Rivers in Cairo, Illinois and watch as thick, mocha-latte and army drab colored waters swirl together in liquid form. Smooth surfaces belie the powerful current that daily sweeps hundreds of thousands of tons of once fertile topsoil down river towards New Orleans. An average of 436,000 tons of soil daily floats suspended within the waters of the Mississippi. This is only suspended sediment. That's the equivalent of 14 Statues of Liberty floating down the Mississippi daily. In living terms, that's the equivalent of about 600,000 cows floating down the Mississippi. Every day.

Beneath the river's surface, heavier farm soils roll and buck along the riverbed heading downstream with the current. Altogether, the Mississippi *"carries about 500 million tons of former U.S. farmland soil into the Gulf of Mexico"* each year. [14] That's about as much soil as the Sahara loses every year. By the square mile, the entire continental United States could easily fit inside the Sahara with some room to spare. The immense volume of Mississippi River erosion is predominately from the U.S. Plains states. Actually though, the scariest part of these statistics isn't the sheer volume of farmable soil being lost, or even that the numbers are increasing. It's that they aren't. The volumes are actually decreasing. Mississippi soil erosion peaked in the 1950's and 1960's, and every subsequent decade has seen a gradual decline. We'd like to think better farming practices have slowed the soil exodus, and to some degree, they have. The larger reality though is that we are running out of farmland topsoils to erode.

There Might Be Hope

The bulk of our prime Midwestern and Plains topsoil left decades ago.

Upriver from Cairo and just north of the Missouri River, lies Divide County, North Dakota. It's a beautiful lush, rolling green prairie. You wouldn't know by looking at it that it's 4 inches away from disaster. When farmers first arrived here around 1900, surveyors found "very black topsoil" 16 inches deep. Most topsoil only extends to a depth of around 6 inches, but the Great Plains were special. They had an exceptional depth of topsoil which enabled this area to produce crops and yields far beyond the scope of ordinary, farmland soil. By 2017 though, the rich, fertile black topsoils of Divide County were altogether gone. Only 4 inches of light gray soils remain. Divide County is fortunate though. Researchers believe other parts of North Dakota have easily lost as much as 19 inches of prime topsoil since 1970 alone.[15] In less than 200 years since teams of horses and mules began to till up the Midwest, parts of America's flyover states are seeing sand dunes pile up in roadside ditches and begin to bury farm fences.

During the Dust Bowl, families, like those of Leptis Magna millennia before, did what they had to in order to survive. For many, that meant packing up and moving elsewhere, generally west. Today, for the first time in history, simply abandoning our farmland to start again somewhere else isn't an option. The ample fertile, virgin farmlands of past generations that enabled the cycle of land use, exhaustion, and abandonment no longer exist.

Most of the earth's surface simply isn't well suited for food crops. Mountainous, rocky, and desertified regions aren't farmable. Today, a scant 10% of earth's farmable land remains uncultivated. Simultaneously, America's prime farmland is rapidly being sealed or capped off for commercial development. Between 1982 and 2007, nearly 35% of prime US farmland was sealed over for urban purposes,[16] and those soils remaining are being increasingly contaminated by heavy metals from air and water pollution that acidify, further desertify, and thus intensify erosion in topsoil.

It's a trend other parts of the world have more history with, and globally, the situation is honestly much more dire. While U.S., European, and South American farm soils are being eroded at rates unrecoverable in our lifetime, some Asian farms are eroding at rates of higher than 30 tons per acre per year.[17] China's Yellow River earned its name by hauling off around 1.6 billion tons of soil every year. Chinese soils aren't just muddying their rivers; they are hanging suspended in the air. The Ganges in India is close. They lose 1.5 billion tons per year. Due to political instability, neither researchers nor the FAO have a clear idea of how much soil sub-Saharan Africa is losing annually, but we do know that productive agricultural fields are fighting a losing battle with encroaching sands from the Sahara increasingly burying their best soils and fields.

Fields generally fail slowly and rarely in unison, but gradual erosion can be the forerunner to sudden, extreme, and apocalyptic-type erosion. During the 1930's Dust Bowl, the U.S. Southern Plains lost an average of 60 tons per acre per year, proving that soil erosion can go from bad to Biblically catastrophic quickly. Windstorms kicked up Plains soils into thick black, advancing walls that enveloped and blanketed towns like a death shroud. Blood rain, a mixture of iron-rich, ruddy dust and precipitation, poured down from the heavens turning normally peaceful rivers and lakes deep red. Abrasive, gritty winds shredded crop leaves, stalks, and produce alike and suffocated entire herds of livestock. Suddenly starved insects and rodents, moving in herds of tens of thousands per acre, invaded cities and towns panicked and desperate to find food. The Dust Bowl seems like an isolated, past-tense event, something that could never happen again, but experts believe it will happen again. Without intervention, we should expect similar episodes with increasing frequency over the next 30 years.

As our best soils erode first, even before we reach bedrock, eroded soils begin to decline in their capacity to grow crops. Healthy crops need fertile topsoil at least 6 inches deep. They may survive in areas with less, but they won't thrive. Vast U.S. land areas can no longer support intense cultivation of grain, vegetable, and fiber crops. Former crop lands may be used for less demanding commodities such as grass, hay, and livestock, but simply converting eroded cropland into pastureland doesn't necessarily protect it from further erosion or rebuild the soil. If a drought hits and hungry livestock graze plants too low, already eroded soils can reach a virtual point of no return. Eroded soils can't absorb water like healthy, fertile soils do. This increases vulnerability to drought and to the incidence of flash floods.

The U.S. government isn't ignorant of these woes. Far from it, every year, they pay farmers close to $2 billion dollars to rest their tired fields from production and help protect these lands from further erosion. The U.S. Conservation Reserve Program pays farmers to set aside roughly 20 million acres a year out of America's almost 900 million acres of agricultural land.[18] This is not because only 2% of American farmland is in serious danger from catastrophic erosion. How much the government has to spend and how badly the market needs food determines the amount of land admitted into the program and the length of time the land stays in the program. CRP lands can rest for a maximum of 10 years but are usually put back into production well before the 10-year mark. This isn't because they've been restored to health. It's because global food demand means even weary, marginally producing lands are worth more in production than being subsidized into rest.

Globally, food shortages are about distribution not lack.[19] In the U.S. alone, 40% of our food is uneaten yearly and finds its way to a landfill. Salvaging even as little as 15% of this would feed up to 25 million more people each year.[20] Extreme weather patterns can cause local shortages, but while poverty exists throughout our world, true famine has been non-existent for centuries. Political conflict, internal barriers to foreign aid, and subsequently rising food prices cause most modern food shortages.

In the coming 30 years though, without dramatic intervention on behalf of our soils, genuine lack will become the norm. How close are we now to famine? Every year, the USDA and the FAO, the UN's agricultural division, inventory the world's known grain stocks, and compare these quantities with the earth's population. They determine how long the world could survive off of currently existing grains if all food production ceased tomorrow. They focus on staple grains such as wheat, rice, and soy due to their caloric density and ability to be transported and stored without refrigeration. In 2022, world grain reserves measured around 70 days.[21] This means, that if a global catastrophe hit tomorrow and all existing crops were destroyed, we have enough food in the global pantry to feed us for about 70 days. The reality behind today's brightly-lit and well-stocked grocery aisles is that our world is living off of about a 2 month food supply. This means that, in the event of a global catastrophe, farmers would have 70 days to successfully grow and harvest a new crop before the world began to starve. Problematically though, in the event of a global catastrophe or series of catastrophes, most wheat varieties require 100 to 250 days from the sowing of seed to harvest. Rice crops need 105-150 days. Soybeans mature far more quickly at 45-65 days, but limited seed reserves and harvesting equipment would be insufficient to make up for a deficit in wheat and rice.

The Bible records Libya's neighbor, ancient Egypt, once experiencing a severe famine that lasted 7 years. Wisely, Pharaoh appointed Israel's son, Joseph, to gather and store 20% of their annual harvests for the 7 years before the famine. Egypt's fields were fertile enough to produce 20% more food than the nation's 1-2 million Egyptians needed for 7 years prior to the famine. When the withering east winds and, undoubtedly, severe erosion they brought arrived, Egypt and her neighbors survived because of these grain reserves. Perhaps from the lingering remembrance of these events and perhaps because the majority of their foods today are imported, the Middle East continues to wisely place a priority on grain reserves, and many Middle Eastern countries keep silos filled with a 6 month or greater grain reserve. China and Russia, two of the world's top grain producers, won't divulge grain reserve amounts, but seem certain to have them. The U.S. does not. America today has a Strategic Petroleum Reserve, but no grain reserves. The U.S. grain reserve program was discontinued in 1996, and America's last grain reserves left the silos in 2008.

In a world where supermarkets' shelves are restocked daily and mountains of produce are regularly pulled from the shelves to spoil uneaten, a false sense of food security prevails. Modern food production and distribution systems have reduced acquiring food from a sun-up to sun-down struggle to a quick stop at the store and a relatively small fraction of our income, yet most of our lost soils are considered to be irreversible and irrecoverable. Without massive and immediate interventions, our remaining topsoil will be destroyed over the next 30 years or less as crop yields decline. Carried off by winds, washed away down rivers and out to sea, this once bountiful and precious resource, at its current, unimpeded rate, will disappear sometime in the next 30 years without dramatic intervention. Exhausted and eroding topsoil isn't the ancient's problem anymore. It's our problem. Like the Romans, the sands of time are literally running out for us. Crops will fail. People will starve, and societies will collapse. To survive, we're going to have to unearth what no previous generation has unearthed. We're going to have to get to the root cause of erosion as no previous generation has done. We are going to have to answer the question of why soils erode.

Chapter 2 Faith Until Sight

Since the 1950's, USDA scientists have led the world in researching soil's physical attributes. Textural and mineral composition, the effects of vegetative cover, weather, tillage, and slope on erosion rates have all been meticulously studied. However, our vast knowledge of soil's physical characteristics and mechanisms never fully answered the question of what causes farmland erosion or how to stop it.

As science advanced through the 19th century, 2 dueling views battled for soil science predominance. Were soils primarily chemical or primarily biological entities? Chemical advocates believed that soils were primarily lifeless, mineral mediums for plant growth that chemicals passed through on their way to root systems. Biological advocates believed that the key to soil fertility would, one day, be found through managing and balancing soil microbes and nourishing beneficial microbial communities. The problem was, in the 1800's, they couldn't find these communities. Nineteenth century laboratory scientist's best efforts to locate and cultivate soil microbes on agar plates and in petri dishes generally proved unimpressive at best and entirely fruitless at worst. This led many to the belief that (lack of)

"...growth on this medium was sufficient to justify the conclusion that the organism did not exist.... All efforts to isolate the specific organisms concerned in the process failed[1]." Farmers and microbiologists often intuitively felt there was more to their soils, especially healthy ones, but evidence was not on their side.

Just prior to WWI, German scientist and future Nazi chemical warfare specialist, Fritz Haber took Carl Bosch's small-scale lab methods of extracting nitrogen from the atmosphere and with the aid of substantial natural gas inputs, replicated the process on an industrial scale. Known as the Haber-Bosch process, its immediate product was ammonia, but the ultimate goal was massive ammunitions production.

Decades later, following WWII, rather than be torn down, former munitions factories remarketed their primary product, ammonia, as a synthetic, inorganic fertilizer. Prior to this, the ability to selectively add massive quantities of an isolated chemical element to our crops had been simply out of reach. Just in time to facilitate the spread and distribution of these new fertilizers came the advent of powerful, hydraulic tractors and farm equipment. The spark to this oxygen and kindling was Norman Borlaug's revolutionary, hybridized seed creations. Immigrating from Mexico to the U.S. and thereafter sweeping around the world, the trio led to exponential increases in agricultural crop production. Agricultural yields exploded. For almost a century, since the USDA began tracking crop yields, the average U.S. corn field had steadily yielded around 25 bushels an acre. From 1950-1959, that average grew to 44 bushels an acre and then 70 in 1960. By the 1990's, corn fields were doling out 123 bushels an acre. Gains in potatoes were even more dramatic. Yields more than tripled from the previous national average of 5,000 lbs. of potatoes per acre to 16,500 in the 1950's, and forty years later, by the '90's, that figure more than doubled to 33,000 lbs. per acre. In a short 50 years,

American fields went from producing an average of 5,000lbs of potatoes to over 33,000 lbs. of potatoes, per acre. Gains in other crops, though perhaps not as dramatic, were substantial. The Green Revolution had definitely arrived. It saved millions from starvation and proved that the purely, chemical approach to the soil brought results. Here was tangible evidence that purely chemical soil manipulation worked. Here was something soil microbiologists lacked.

Alongside these technological advancements, evolutions in finance along with increasing land prices and equipment costs, birthed the consolidation and specialization of formerly diversified operations on increasingly massive scales. Survival for many agriculturists meant taking a page from the assembly line. Small-scale, family farmers lacking the volume of land necessary to produce competitive volumes began falling by the wayside. Survival in an already risky endeavor virtually mandated large acreage, monoculture or single crop farming. Farms increasingly looked to new government crop insurance and subsidies as potential bailouts to assuage the additional risk and debt they assumed in transitioning from traditional diversified operations to potentially more lucrative monoculture operations.

With this increasing specialization of agricultural operations, the traditional animal and plant agriculture cycle that recycled local manure by applying it to crops disappeared. Livestock production began to also be consolidated into Confined Animal Feeding Operations or CAFOs where livestock waste that formerly had been an essential crop nutrient was, likely for the first time in human history, more problem than blessing.

By the 1950's, the inherent difficulties and often fruitlessness of soil microbiology research combined with the immergence of patentable and hence far more lucrative means of addressing soil fertility gave root to the idea that soil microbes were probably of little value and their utter extermination, a boon.

> "The protozoa were looked upon merely as 'injurious forms' or as the 'enemies'... The fungi were considered either nuisances or 'dust contaminants'...None of these organisms, however, were given sufficient consideration in any systemic study of the soil population; if they were considered at all." ~ Selman Waksman, Noble Prize Winner. [2]

Somewhere along the way, these presuppositions began to be taught as established and then static truth to generations of farmers by chemical salesmen and the agricultural colleges whose research projects were often financed by these same chemical companies. Over the 20th century, the perception that agricultural soils were by and large, a purely mineral medium and devoid of any worthwhile life became familiar, and then comfortable.

Along the way, a curious phenomenon began to develop. Crop pest infestations, fungi, and viruses previously unknown overtook and destroyed formerly healthy fields[3]. Simply maintaining previous yields required increasing amounts of these synthetic, inorganic fertilizers and increasing amounts and new formulations of pesticides to combat plant infestations and diseases. While these synthetic chemical applications increased, erosion was accelerating. Soil failures broadened and soon, we found ourselves standing at the confluence of the Mississippi and Ohio Rivers watching as turbulent river waters hourly packed up hundreds of tons of once rich and fertile farmland topsoil and shipped them off towards the Gulf of Mexico. The darker side of the Green Revolution was troubling.

All the while, a small remnant of soil microbiologists tenaciously clung to their beliefs that soil fertility sprang from soil life. For decades, they toiled away with much theory and little fruit. In the late

1980's and 90's, they suddenly gained access to high-powered microscopes that had previously only been available to medical researchers. These microscopes proved that much of what had previously been literally dumped into the laboratory trash can was a veritable goldmine of soil fertility and, even more wonderfully, of human nutrition. Previously unexplored magnifications brilliantly illuminated amazingly vast, dense, and complex soil ecosystems. The long elusive proof had been unearthed, and it came in unimaginable quantities that, like the ocean's depths, we are far from being anywhere close to having fully explored them today. Fertile, healthy soil is not a lifeless, chemical medium occasionally frequented by enemy insects intent upon destroying our food supply. It's inhabited and even densely saturated by more life than we could ever have imagined! Trillions of soil life forms can easily exist in just half a teaspoon of healthy soil. We may have identified as little as 13% of soil's microbial organisms, and of those identified, many if not most are not well-understood."[4]

Elementary students have long been taught that plants can make all the food they need with only sunlight and water. False. This is entirely false. In ancient times, plants are believed to have been entirely dependent upon a form of fungi known as Mycorrhizae for sustenance. While many crops today generally survive either without or with minimal Mycorrhizae, they still thrive best when living in community with them. The Mycorrhizae and host plant relationship is a bit like a marriage. Like a good husband, Mycorrhizae diligently provide for and protect the plant, and the plant, likewise, responds, like a wife, by feeding the Mycorrhizae the carbs it cooks up and lots of them.

Mycorrhizae grow as microscopic fibers out of the plant's root and root hairs branching out like veins, into nutrient, superhighways that crisscross and create a vast nutrient transportation network. In the process, they anchor the soil in place like woven mat and extend the root systems reach far beyond the plant's own abilities. Individually 1/25th the diameter of a human hair[5], mycorrhizae fungus fibers are normally invisible to the human eye. Their size allows them to delve into rockier soil areas otherwise unnavigable to exponentially larger root hairs and roots. More than a mile of a single strand could fit inside a thimble. The creeping process of rock weathering, by wind and rain, was formerly believed to break minerals down into small enough particles for plants to absorb. Today, we know that isn't true. Plants cannot convert inorganic soil minerals and nutrients into useable forms, and they are limited to what is immediately available in the soil around them, unless Mycorrhizae help them. All along its strands, Mycorrhizae exude powerful, carbon-rich enzymes into the soil. These enzymes saturate nearby inorganic minerals and nutrients slowly drenching and dissolving otherwise bound and inaccessible soil minerals and nutrients into forms that plants can absorb and digest[6]. The fungi then "brings home the bacon" and transports these now digestible mineral forms back to their plant partner. Mycorrhizae supported plants contain significantly, higher amounts of vital nutrients such as zinc, iron, calcium, magnesium, manganese, and phosphorus than those without.[7,8,9] They're naturally higher in nitrogen also, an essential building block for all protein.

Should a harmful nematode or enemy worm be headed in the direction of its plant, the Mycorrhizae will flex the tip ends of their fibers into a loop, creating a snare in the path of the approaching nematode. As the unsuspecting nematode worms its way through the loop, the snare suddenly tightens, lassoing the nematode, and holding it snugly captive. The Mycorrhizae then sends out digestive enzymes that first dissolve and then devour the nematode from the inside out. Adhesive enzymes are another tool in the Mycorrhizae arsenal.

Harmful nematodes must only brush up against the Mycorrhizae for it to release these enzymes. Once contact is made, the nematode cannot break away from these incredible adhesives. Powerful digestive enzymes will then pass through the point of contact with the nematode and begin to digest it from the inside out. These predigested nutrients are then sent back as ready-made, protein-rich, food sources to the host plant. Without the Mycorrhizae to protect its spouse, the harmful nematode would likely do serious damage to the plant, but when paired with the Mycorrhizae, the nematode becomes an essential complex protein food source for the plant and its offspring.

Another top priority for the Mycorrhizae is constructing a healthy barrier between their host plant and plant predators within the soil. It does so by building a hard, protective coating over the outer root cells. Besides defending its host plant against harmful denizens of the soil, Mycorrhizae produces proteins that protect its host plant from harmful genetic changes and damaging heavy metals within the soil[10]. Mycorrhizae also defend their host plant against a variety of fungal root diseases by producing their own special antibiotics that guard the host plant from infection.

While the Mycorrhizae husband is protecting and providing for the plant, the plant is busy taking those groceries and using them along with the carbohydrates they synthesize from the air and sun to cook up a nutrient dense meal for the Mycorrhizae. After the Mycorrhizae digest the complex carbs, they exude needed carbon reserves back into the surrounding soil in the form of phytolipids or fats and an amazing, complex, organic protein known as Glomalin.[11,12] Like gazillions of microscopic grain silos below the soil, Glomalin also serves as a storage mechanism for perhaps as much as one third of all the world's carbon. These stored carbon reserves can last for years until plants need them.

Glomalin also loosens up hardened, compacted soils and creates vital oxygen pathways for our subterranean cities to breathe more easily. Far from a lifeless, mineral medium, healthy, living soil breathes. For many decades, we've known that our forests and oceans breathed, but the collective inhaling and exhaling of healthy, fertile soil microbes has been largely ignored along with the infrastructure that makes it possible. As these microscopic soil civilizations exhale carbon dioxide, plants receive it, breathing in deeply. Plants reciprocally exhale oxygen, and the soil microbiome inhales it, gratefully. Each has a distinct role; each needs the other.

The beauty of this relationship is particularly evident under drought stress. When stressed due to drought or extreme temperatures, plants will willing sacrifice their own growth to feed more carbohydrates to the Mycorrhizae in an effort to strengthen the Mycorrhizae. They trust the Mycorrhizae to retrieve more soil minerals and nutrients for the starving plant that will help it through stressful times. This beautiful partnership also ensures that host plants are less likely to experience drought or stress in the first place. When rainfall is adequate, Mycorrhizae store water and slowly release it to their host plant acting as a vast multitude of tiny aquifers conserving and supplying water according to their host's needs. Like a system of mini-Yetis up and down the roots, these water stores help shelter plants during extreme heat or cold by serving as insulation which respectively cool and warm root systems. All of this gives plants colonized with Mycorrhiza a competitive edge over plants without Mycorrhizae. The Mycorrhiza provides its host plant with more nutrients than the lonely, unwed weed can secure, resulting in faster growth and monopolization of nutrient, light, and water resources. This is why wilderness areas visibly withstand drought far better and longer than neighboring cultivated fields with fewer Mycorrhizae resources.

Amazingly, Mycorrhizae are nonmonophyletic, meaning they haven't evolved from a common genetic ancestor. How do we classify and organize a species that doesn't genetically resemble the family we've placed it in? In this case, we group and classify them by how they reproduce. Many experts believe that up to 90% of all plant species are designed to host some form of Mycorrhizae and did when the earth was new. This includes crops. In 2007, around 150 species of Mycorrhizae were known, and today, we're at over 342[13] and climbing. For these reasons, researchers anticipate that the 342 currently known species are a mere tip of the iceberg.

These nutritional rockstars are just one type of soil fungi, and most soil fungi are still unknown. One study over about 3.5 acres of Swiss farmland recorded 2,554 types of fungi. Almost 2,000 of these, we know too little about to classify them much beyond simply being fungi. The same 2014 study, using the same samples, measured 3,877 types of soil bacteria units. Of these, scientists could only positively identify 61 species.[14] Each has a specific function and a specific purpose.

This was beautifully illustrated by a 2017 study in which scientists sought to identify the types of bacteria and fungi that grew inside of wheat plants. You read that right. These are beneficial, native bacteria and fungi that are essential to plant growth and live inside the plant. Eighteen unique fungal communities and 20 bacterial communities grew within these wheat plants from seedling to harvest. Like a symphony, different fungi and bacteria emerged and catalyzed plant growth at each growth stage. Within these communities, over 31 million distinct bacterial and fungal RNA sequences were counted. The vast majority of these could not be identified any further than that they were fungal or bacterial and previously unknown.[15] Pure sunlight and water alone weren't causing these seeds to germinate; it was an enormous and diverse symphony of fungal and bacteria communities that supported and perhaps even informed the seed, then the seedling, and lastly the mature plant of when the proper time was for the next stage of life.

Yet another study found up to 6,000 unique bacterial genomes have been found in just one gram of soil.[16] Still another 2014 study comparing soil bacteria in Iowan corn fields with Iowan prairies found 5.1 million bacterial DNA sequences in soil samples. To quote the study, the volume of genetic data recovered from these soil bacterial genomes was so immense that, *"a single metagenomic project can readily generate as much or more data than is in the global reference database."*[17] Only a handful of these studies have been carried out, but it will take some time to glean through them.

Soil bacteria themselves, for the little we know about them, are quite amazing. If you've ever caught the beautiful, earthy aroma of freshly tilled, rich, healthy soil, then you've smelled the bacteria Actinomycetes. For decades, researchers struggled to classify Actinomycetes which had characteristics of both bacteria and fungi. The dark brown and black pigments often associated with healthy soils are also produced by Actinomycetes.[18] Probably the most important bacteria you've never heard of, this family of soil bacteria have played and continue to play a vital role in human history for their intelligent anti-cancer, anti-bacterial, anti-fungal, anti-viral, and enzymatic properties[19]. Popular antibiotics such as Neomycin, Tetracycline, Streptomycin, and Actinomycin all originated in soil bacteria. So did Ivermectin. Today, over 5,000 compounds, up to 90% of commercially available antibiotics, and more than 70 individual antibiotics can hail some part of their origins back to these highly, complex soil microbes.[20][21]

Like Mycorrhizae, some Actinomycetes species will inhabit a host plant and then produce natural plant protective antibiotics from inside of the plant to defend it and likely those who consume it from infection. We still don't entirely understand why these naturally occurring antibiotics are in our soil. They are the most genetically complex bacteria on the planet. *Streptomyces coelicolor*, for example, has a genetic code with 8,667,507 base chromosomes, the largest number of genes ever discovered in a bacterium[22], any bacterium. Within these genes are more than 20 clusters that govern unknown or secondary metabolites, and *"an unprecented proportion of regulatory genes."* Researchers hypothesize that most of these regulatory genes are *"likely to be involved in responses to external stimuli and stresses."* In other words, these millions of genes inside of bacteria likely serve to guard and protect their plants. Evidence suggests that they produce plant growth promotants and inhibit harmful microbes and diseases better than manmade fungicides.

As human antibiotic resistance grows, researchers are heading to increasingly remote areas to search for new Actinomycetes species. Man-eating Bengal tigers, King cobras, saltwater crocodiles, pirates, and an endless labyrinth of gray, clay estuarine channels shaded by thick, mangrove jungle doesn't sound like the ideal soil microbiology lab, but it's in the Sundarbans of eastern India and other untouched soils that medical researchers are searching for previously undiscovered species of Actinomycetes. Pharmacologic companies are sending Actinomycete search teams across Eastern India, Bangladesh, Indonesia, the Mongolian steppes, and even near the Temple Mound in Jerusalem. Why does this incredible abundance and diversity of soil antibiotics exist in the first place is an almost baffling question. Like many bacteria, Actinomycetes certainly play a vital role in decomposing decaying plants and manures and converting atmospheric nitrogen into usable plant forms. Could they also be inherently healing and protective of us, the plant consumers?

Another of soil's essential citizens is algae, and like much of the soil microbiome, we know little about it, other than its existence.

Because soil algae growth relies on sunlight, it only grows on the soil's surface, but its work is powerful in helping soil particles stick together. This creates natural soil agglomerates that open-up air spaces and increase air flow for the soil microbiome below to breathe. Algae nourishes the soil by pulling in atmospheric nitrogen and building it into protein forms the soil microbiome can potentially use. What's especially fascinating is that soil algae seem to flourish even in desert soils with sparse soil microbes.

Dead plants, manures, and animals---the end of life above the soil marks the beginning of soil's life cycle below it. Decaying organic matter attracts ever insatiable soil scavengers--bacteria, fungi, and microbes. Shortly afterwards, these scavengers expel these nutrients. They are reconstituted in the form of complex organic molecules known collectively as humic substances or soil acids. In the soil microbiome, nothing is wasted. These humic substances are nutritionally dense soil pantries that provide healthy meals for both bacterial and fungal grazers, the herbivores and omnivores of the soil. An excellent marker of soil health, soil acids provide a benchmark of available food reserves for the soil microbiome. As such, they are foundational to topsoil's creation, sustenance, and rejuvenation. Healthy topsoils produce soil acids continuously, but because the process is bacterially driven, they do so fastest in warmer climates and weather. In Antarctica, humic and fulvic acid production slows to a near halt. In tropical climates along the equator, soil acid production can occur swiftly. Like most of life though, the healthiest soil civilizations tend to thrive best within moderate temperatures rather than extremes.

As scientists are just beginning to understand soil acid production, they've been faced with the challenge of classifying and naming these amazing substances that don't fit into any previously recognized categories. Soil acids aren't animal, vegetable, or mineral.

They are carbon, oxygen, and hydrogen clustered together with a wide variety of other elements to create enormous groups of DNA, amino acids, and carbohydrates of immense size. The size of a soil acid when, compared with most, previously known chemicals, has been likened to comparing a salmon to a whale. Poly just means many, and that's what these soil acids are believed to be. Poly-what is the question though.[23] Like snowflakes, their individual molecular structures are unique from even their nearest neighbors and vary greatly in both complexity and diversity. This diversity is essential to healthy, living topsoil, and yet they are simply too diverse and too complex to soundly fit in any currently existing chemical or biological category. For the time being, we call them soil acids. We can group them by color, by temperature preferences, and by carbon content, but everyone knows the term soil acids is just a vague colloquialism we're using as a place holder for lack of a better understanding. Adding to the complexity, many of the bacteria, fungi, protists, and nematodes that create these soil acids haven't been identified yet.

In wine, cheese, or preserved meat production, differing types of bacteria and nutrient levels allow for nearly limitless varieties of scrumptious flavors, enticing shades, and appetizing aromas. Soil acid production is no different. The amount and diversity of nutrients provided by decomposing organic matter combined with the quantity and diversity of the soil microbial species determine the amount, types of nutrients, and even the color soil acids lend to their home soils. Coppery gold soils may be rich in fulvic acids while humic acids tend to turn the soil a rich, dark brown.

Soils acids are the soil microbiome's first choice buffet. They also like Mycorrhizae will, over time, slowly dissolve the minerals and rocks surrounding into plant absorbable forms. In times of abundant rainfall, soil acids absorb water into their structures, transforming and expanding from solid to gelatinous substances that can hold up to

98% of their weight in water. Like Mycorrhizae, they too become widespread, mini aquifers that in drought or extreme temperatures provide plants with both an additional reservoir of water and a natural means of cooling or heating root systems. Under heavy rains, water-saturated soil acids can link together with other soil particles to form an invisible, water-resistant fabric on the soil's surface which helps to resist erosion from damaging winds and rains. When combined with plant cover and Mycorrhizae netting, soil acids, the raincoat of the soil, provide serious defense against soil erosion.

During their short lives, soil microbes have the potential to reproduce and multiply with incredible speed. To an extent, this is a blessing. When soil fungi and bacteria die, their bodies release stored organic nitrogen, in the form of complex proteins and amino acids in addition to other nutrients back into the soil, further feeding the soil microbiome and the plants it partners with. Were it not for soil predators though, soil bacteria and fungi might completely devour the world in short order. Thankfully, healthy soil microbiomes have checks and balances in the form of soil protists and nematodes, the bacterivores and natural predators of the soil.

No one really knows what a protist is, and that's part of their charm. Invisible without extreme magnification, these peculiar living creatures called protists are neither animal nor plant nor fungus. Despite our laboratory advances in preserving soil microbes, protists continue to be a challenge to extract from farmland soils. They are delicate, ornate, and jewel-like and yet strong. Like Mycorrhizae, they're mostly nonmonophyletic and did not genetically evolve from a common ancestor, once more complicating our studies of them. Unlike Mycorrhizae, they're classified by how they move rather than reproduce. Generally single-celled, they ooze or flutter, depending upon their constitution, through the soil riding upon steeds of water molecules to wage war upon other protists and feast on bacteria. One such group of individuals is the naked amoeba, whose gelatinous blob of a body, bizarrely, can, sans muscles, senses, or nervous system, chase down, engulf, and consume its prey. Some refer to this as a hunting style. Others call it a *"grazing aptitude,"* but regardless of whether you see them as more lion or more lamblike, their role is essential to a healthy, functional soil microbiome.

Biofilms are sticky, slimy substances that some bacteria produce and shield themselves with as a protective, defense mechanism. Recently, they have come to the forefront of human medicine as potential root causes for a variety of bacterial diseases. Some protists seem to have been specifically designed with equally protective exteriors that allow them to wade unharmed into those biofilms and graze on these bacteria. Like Velociraptors or birds of prey, raptorial protists claw, grasp, and tear apart bacteria and other protozoa as they devour them. Heliozoan protists look, as their name implies, like mini-suns that wrap their bacterial prey up in happy, sunbeam-like extensions before absorbing them.

Nematodes are microscopic worms. A couple of billion of them could easily fit into a teacup, but you won't find anywhere near that amount in healthy topsoil. Within a square yard of healthy, farmland soil, you'd likely only find a couple million. Like soil bacteria and fungi, nematodes are another tremendously diverse group. There are probably as many as 1,000,000 distinct species of nematodes. Contrast this with the little more than 5,000 unique species of mammals in existence or perhaps the 10,000 species of reptiles for a sense of perspective. Until recently, nematodes were synonymous with massive crop root destruction, and the only good nematode in decades past was considered a dead one. Yet many, if not most, nematodes don't eat plants. They eat soil bacteria and fungi and lots of it. Some are even equipped with a straw-like mechanism that can slurp up bacteria 20 or more at a time.[24] A 2017 study of German, wheat farm soils found the nematodes in their soil samples fell into 5 distinct groups, plant eaters, bacterial eaters, fungal eaters, predators, and omnivores. A little more than half of the nematodes in their samples ate primarily soil bacteria and fungi. A little less than half dined upon exclusively on plant roots, and unsurprisingly, very few, only 1.3% were either omnivores or predators.[25] Even in healthy above ground ecosystems, such as prairies or forests, predators comprise a small percentage of the total number of creatures. Too many predators within the soil would decimate prey species. Too few predators and prey species would overrun and destroy available vegetation leading to the prey species' ultimate starvation. Nematodes are essential to healthy soil and ensure soil microbiome balance.

As we peer into the microscope, a miniature, subterranean ecosystem that mirrors those seen in our forests and prairies sharpens into focus. Every species, even predatory species, are absolutely essential.

In 1926, the last wolf of Yellowstone National Park was killed, and the species was considered to have been completely eradicated.[26] From a pioneer mindset, it made sense to prioritize our own survival and that of a key food source by eliminating what was considered to be a dangerous predator, but there were other, unexpected consequences. The park began eroding down the river. Park river banks were literally washing away. Rangers thought they were preserving the elk herds by eliminating the wolves, but pressure from the wolves had kept the elk moving along and from grazing any one area too intensively. Without the wolves, the elk overgrazed young willow and aspen trees whose roots and root partners had helped to hold soil in place along the river banks. For the next 75 years, the cascading damage continued and even ultimately shifted the course of the park's rivers. It became apparent that an entire ecosystem was dying, and all because of the removal of just one species. Despite protests and fears that they would devour all of the park's elk, a small pack of just 14 Canadian wolves were introduced to the park in 1995. When the wolves returned, so did the formerly bounteous ecosystem. Willow and Aspen trees grew again. The songbirds and beavers returned. River bank soils were stabilized. Balance is essential.

In addition to the carbohydrates and fatty acids their plant wife cooks up and feeds them, Mycorrhizae hunt down and consume nematodes. Without nematodes, Mycorrhizae lose key, complex protein and fatty acid components of their diet that they can no longer share with their plant wife. The plant wife, in turn, can no longer provide these complex proteins and fatty acids to animal or human consumers.

Like wolves, bears, sharks and other predators that were formerly demonized but now have been recognized as essential to ecological health, nematodes are starting to be recognized for their critical role in soil health. Whether above or below ground, predator species keep herbivore species and vegetation in balance, and soil bacterivores ultimately strengthen soil bacterial populations by picking off the weak and the stragglers. Balance is as highly prized within the soil microbiome as it is in forests or prairies. Decimation of any one species creates imbalance within the entire soil food web. Every community member is essential.

Predators and prey, omnivores, herbivores, and carnivores—regardless of how inconsequential one of these miniscule individuals seems, the collective health of the entire subterranean microbiome depends upon these individuals doing their jobs. To provide a diverse and rich array of nutrient resources for themselves, for their plant wives, and for plant consumers, each of these communities must be present and in-balance with one another. It is these orderly and balanced communities whose industrious citizens build and maintain stable soil infrastructure, a defense system, and a food and water supply that benefits all communities.

In a thriving soil microbiome, its tiny citizens are ever growing, producing, reproducing, being eaten alive or dying, and ultimately are recycled back into the soil microbiome at an incredible speed. A crescendo of complex nutrition is fabricated together as nutrients are developed and transferred from decomposing materials to bacteria to nematode and protist to Mycorrhizae to plant to plant consumer. Nothing is wasted. Even when soil microbial bodies die unconsumed, their bodies disintegrate into organic proteins, complex carbons, and fatty acids for other members of the soil's food chain to consume. Each individual within the greater collective community plays a vital role in the soil microbiome, and without one of these elements, soil's ability to nourish plants and ultimately people is diminished. When too much unchecked power or advantage is given to one population or another, a soil microbiome out of balance quickly becomes a fierce, tribal wasteland.

Soil science once seemed a bit like geology where we could sort soil like rocks by their colors and textures. We could measure soil particle size and texture. We could weigh soils. We could carefully procure soils samples, put them in a petri dish, and examine them under a microscope. We could only see a barren wasteland. It turns out, we simply weren't looking closely enough. Now we see that soil science is more akin to marine biology than geology. Soil microbial species are simply too dynamic and too sensitive to be observed by yesteryear's laboratory and research techniques.

Farm soils can never again be seen as a mere mineral medium and chemical conduit. We've come full circle to the agriculturists our forefathers were who intrinsically knew that fertile soils were teeming with rich, abundant, and healing life, even if they couldn't prove it. It is that soil life that is essential both to the health of the plants grown within those soils and the animals and people who consume them.

When we return to the desolate shores of Leptis Magna, when we see the sands of the Sahara strewn about in the atmosphere above U.S. farm soils, when we stand on the banks of the Mississippi and watch the turbid waters carrying American farmland off to the Gulf of Mexico, the question is obvious. If the soil microbiome is the soil's inherent natural defense against erosion, why isn't it doing its job?

Chapter 3 Native Farmers

"The oxen were plowing and the donkeys feeding beside them, when the Sabeans raided them and took them away-indeed they have killed the servants with the edge of the sword, and I alone have escaped to tell you!" ~Job 1:14-15

Arable: To plow; Latin. capable of producing crops; suitable for farming; suited to the plow and for tillage...

Yee. Haw. Today, these words may bring to mind clownish, frolicsome country themed events where cheap, straw hats, hayrides, face-painting, and flannel abound. In ages past, they represented teamwork, hard labor, and sustenance. "Yee", likely a corrupted form of "Gee", translates to "Go right." "Haw" commanded a plow animal to "Go left." These directional terms were spoken by a plowman to guide his draft team. Variations of these terms are thought to date back to at least the 1500's and likely well beyond, but plowing, as a practice, appears to have grown up alongside the foundations of human civilization. Commonly thought to date back to at least 2000 B.C., early plowing references frequently appear in ancient, Egyptian artwork and written Biblical accounts such as Job losing his 500 yoke of oxen to the Sabean raiders. The donkeys grazing beside Job's plowing oxen were likely an agricultural technique used to trim down tall grasses and clear the fields for easier plowing.

From antiquity until the 1900's, plowing was an intensely physical task and generally limited to a double yolk of heavy-boned, sure-footed oxen followed by a plowman. Wealthier operations might have had the benefit of a mule team or even draft horses. For millenia, oxen, mules, and horses were specifically bred for this purpose. Stout-boned and muscular, these heavy lifters or gentle giants of the livestock world listen attentively with willing hearts for the verbal commands "Step-up. Gee. Haw. Whoa." In spite of the work's fatiguing nature, most draft animals seem to both genuinely enjoy their jobs and the camaraderie that comes with it. The reins often lie slack as they pull forward eagerly on their own at a brisk walk. While trusting and relying on his team, the man behind the plow must be quick to keep up even while finding his own way over both unbroken and freshly turned earth. Upper body and core strength are tested as he rights, maneuvers, and steadies a heavy, 100-170lbs. of plow that often wants to swerve, pitch, and dive away from the unbroken ground back into the last furrow.

A pragmatic physicist, an accurate mathematician, and a meticulous mechanic, the skilled plowman deftly applies his knowledge and understanding to a variety of soil and weather conditions. Like a captain guiding his crew to harness the wind as he steers the ship's rudder, a good farmer harnesses the strength of his team while overseeing the rudder of the soil, the plow. The results appear simple. The plow slices through the soil, lifts it up, and folds the surface layer over and under. Like folding an omelet, the end result of a well-plowed field ought to display straight, smooth lines. The straighter the row, the more space efficient a crop is, and the greater the potential yield. Poor weather or soil conditions, an inexperienced or poorly led team, ill-calibrated or ill-fitting equipment—any of these results in a field that more closely resembles a plate of poorly mixed, scrambled eggs.

Prior to the advent of tractors, a furrow, the space between plant rows, was just wide enough for a single plow horse to walk down, and most plows created 1 or 2 furrows at a pass. At utmost, furrows ran 200 yards in length. This was the maximum distance a team could comfortably pull a plow, in ideal soil, before needing a break to catch their breath. In average to poor conditions, furrows held closer to 100 yards long or required a stronger, larger team. A full day's plowing for a solid team, in good conditions, meant plowing an American football field or a little more than an acre. This is where the term horsepower hails from. One horse equals one horsepower, the equivalent workload a single, draft horse can comfortably pull in a day.

It's said that two well-worked acres could comfortably feed a family of five, and an eight-acre homestead was about the maximum a man would want to tackle on his own. When John Deere's revolutionary steel plow debuted in 1837, it's unique curvature and glossy, smooth surface empowered farmers to cut through thick, sticky prairie soils with previously unimaginable ease and speed. It was a tremendous step forward, but barely a century later, the emergence of tractor and hydraulic power would replace draft livestock altogether. The acre of plowing that once took a team all day now takes as little as about 5 minutes. Literally.

It might be difficult to overestimate the sheer physical relief that the invention of the tractor brought to modern agriculture. The team's strength and the prowess of the man behind the plow to keep it upright once determined depth and the evenness of rows. Today, plowing depth can be easily adjusted from a touch-sensitive computer screen mounted next to an air-ride-equipped, lumbar-supported, leather seat within an air-conditioned cab. LED lighting, blue-tooth, 4G, online tractor diagnostics, and a farm GPS guidance system offer further comfort and support. That is if a farmer can afford them. Tractor sticker shock is real. Fresh off the lot, a new farm tractor can easily run double the cost of the average American home. Regardless, agriculture's traditional constraints of a team's physical exertion and available daylight don't limit the size of a field anymore.

The invention of the tractor didn't really make an inherently difficult job easy. It simply made accomplishing it on a much larger scale possible. On the one hand, it meant fewer farmers could feed more people than ever before, and on the other hand, meant that decreased food prices forced farmers to produce greater and greater yields to survive themselves. Producing greater yields required increasingly larger and more powerful equipment.

Early tractors meant an immediate difference between the strength of 2 horses requiring care, feeding, and shoeing and the horsepower of 15-25 horses that didn't require a pasture of their own, pulling in unison. The tradeoff came in seasonal machine maintenance--greasing bearings, changing belts, changing oil, and airing up tires. While a major improvement in terms of workload, these early tractors were a far cry from today's agricultural tractors which boast 200-600 horsepower. Imagine 600 draft horses pulling in unison across a field, and you've got the modern, agricultural tractor.

Along with increased power came increased weight. Altogether, a team of draft horses or oxen with complete harness, plow, and plowman might weigh in around 5,000lbs. Contrast this with today's closer to 40,000 lbs. of precision agricultural equipment pressing down upon the earth. Tack on a plow or other implements, and another 10,000-20,000lbs. gets added into the mix.

It's in part the craftsmanship that thriving, subterranean, microbial civilizations pour into their infrastructure and airways that enables them to survive below the surface, and in part, it's their sheer proximity to the surface that enables them to access the oxygen they need. Most of soil life dwells in the upper 6 inches of the soil, the topsoil, but like most living creatures and their dwellings, topsoil simply wasn't built to withstand a concentrated 50,000-60,000 lbs. repeatedly rolling over it. If you've ever been stepped on a horse, you've likely lived to tell your painful tale, but what about being run over by a tractor? Every pass of these modern beasts of burden gradually weakens and ultimately collapses subterranean infrastructure suffocating and crushing the air-breathing, microbial life within it.

Imagine a woman in Sub-Saharan Africa walking to the well to gather water for her family. Through much practice and necessity, she deftly balances a 5-gallon Jerry can of water on her head. It's challenging, but possible. Now, imagine the same woman balancing 50 times that load vertically on her head. This is what we've been asking our soils to do under the weight of increasing machinery. Between plowing, planting, fertilizing, and harvesting, making 12-15 passes over a field in a season with a tractor and various implements is not uncommon. Farm soils are increasingly pulverized into more of a road base than a soil microbial nursery[1]. It's a vicious cycle. As the soil becomes increasingly compacted, the response is generally more plowing to recreate the sub-surface airways that plants rely on to breathe and that soil microbes intuitively build.

Plowing is probably the most common form of tilling, but there are many forms of tilling. The common goal of all tilling is preparing the soil for planting by killing competing weeds and breaking up the soil. Why we till is simple, but the best methods for tilling remain as up for debate today as they have since antiquity. *"...the ancient Greeks, who plowed first and then sowed were quite amazed at the*

Egyptians, who carried out planting by reverse fashion.[2]" By breaking up the soil, young seedling roots were thought to be able to burrow down into it more easily. To this end, tilling was also believed to improve tilth, the silkiness or "feel" of a soil. The greater the tilth, the more freely soil slipped between your fingers. Back when soil was believed to be a purely lifeless, mineral medium and chemical conduit, this all made perfect sense or seemed to. The soil needed our help. It had no natural, God-given mechanism to create soil infrastructure that would bring oxygen to plant roots and allow them to grow well. Only, it did, and it does. The soil is not a lifeless, mineral medium. Tilling tines or plow blades slicing through the ground, do not simply smooth out soil consistency and uproot competing plants; they wreck soil microbiomes, the same microbiomes that genuinely build tilth. Like a ground-splitting earthquake preceding an avalanche of soil and plant debris, tilling rips apart microbial communities and infrastructure while collapsing soil microbial airways, burying air-breathing microbes alive, and often suffocating them under 4 to 12 inches of soil chaos.

Among the wreckage lie broken strands of Mycorrhizae. Like most fungi, Mycorrhizae reproduce through spores. Unlike fungi above ground though, Mycorrhizae spores aren't carried along by wind or water. Traveling below ground, they meander from root-to-root or hitch a ride with an earthworm until they match and marry the plant of their dreams. A few species of Mycorrhizae grow well with many different varieties of plants, but nearly every species of Mycorrhizae has one family of plants that they're partial to. Once the Mycorrhizae and its plant are wed, they grow and mature together for a year before producing spores of their own. Herein lies the trouble. Most major food crops like wheat, corn, and soy aren't rooted in the ground for a year. They are rooted for a season. At the season's end, they are harvested, and the remains are often plowed under thus ending the union of Mycorrhizae and plant before the next generation of spores can be sent off.

When oxygenated muscle was the limiting factor in furrow length, crop roots were closer to field borders. Unplowed, field-framing borders hosted vibrant nurseries brimming with healthy and diverse soil microbiomes that were able to send out fungal spores, bacteria, and microbes to replenish nearby depleted, crop fields. Even if the crops themselves were too short-lived for their own Mycorrhizae to reproduce, spores were still close enough to repopulate new crop roots. The sheer, uninterrupted expanses of today's farm fields, however, means that after tilling shreds and crushes existing Mycorrhizae, undisturbed, healthy, soil microbiomes are too far away to effectively restock these fields. Healthy, fertile microbial nurseries are being pushed into smaller and smaller reservoirs or microbiome preserves. Over time, between the continued cycle of Mycorrhizae's inability to reach reproductive maturity and the destruction of existing colonies through heavy tilling, many species of Mycorrhizae are driven to extinction across our farm fields. [3]

Amazingly, some Mycorrhizae species from the family Glomeraceae may survive plowing. These individuals have the unique ability to reconnect and heal broken fragments when nearby. They can even birth spores from them. Of the 342 currently known Mycorrhizae species though, these are the exception, and despite Glomeraceae's resilience, many farm fields have no Mycorrhizae at all.

The Mycorrhizae dating game is a tricky one. Not every plant is attracted to every Mycorrhizae species. Scientists don't yet understand the mechanisms behind these attractions or preferences, but when these matches blossom and thrive, the compatibility is obvious. Happy Mycorrhizae marriages results in far healthier plants and dramatic yield increases. The recent flood of new Mycorrhizae species discoveries haunts us with the question of how many unique, precious

Mycorrhizae species intensive farming methods may have unknowingly and unintentionally obliterated throughout human history. Certainly, at one time, the lands of ancient Mesopotamia, ancient Egypt, and ancient Leptis Magna were propelled towards greatness, in part, due to Mycorrhizae. Were these species the same Glomeraceae we find in farm fields today, or were there others that feed their landlords richly before falling victim to overzealous plowing? Conversely, how often throughout the millennia has man cursed the inherit difficulty of plowing the land without realizing that his natural limitations of physical strength were sheltering our very survival?

Homeostasis--it's a particular pH range, a particular pressure, and a certain level of oxygen saturation. The human body knows this continuity and equilibrium are essential to life. So does the soil. Fungal communities, bacterial communities, and protist and nematodes have an amazing ability to self-regulate their own metabolism, and act as a buffering system to protect plants and themselves from smaller changes or imbalances in the soil's metabolism. They will even voluntarily shut down the growth of their own community, at times, to help maintain this homeostasis. Like our own bodies though, there are limits to this, and systems can become overwhelmed by external forces. Short-term imbalances may be temporarily endured and repaired, but long-term imbalances lead to illness and ultimately death for soils. Included in this homeostasis is the regulation of soil protein and carbon. Soil's glue, safety net, and kitchen pantry, carbon is the singular element that all living creatures share in common. The study of organic chemistry is the study of carbon life forms. Healthy soil microbiomes have diversified sources and stores of carbon just as they have checks and balances between bacteria and fungi. These carbon reserves are also the primary building blocks and food sources for all life and especially the soil microbiome, the genesis point for all nutrition. For the soil microbiome, carbon is the bread of life. Man may succeed on a low carb diet, but the soil microbiome does not. Carbon is often the limiting factor in soil microbiome growth and diversity. Healthy, fertile farm soils maintain a carbon to protein ratio of about 12:1, and they do so, in part, by regulating protein production among a diverse variety of soil microbial communities. Carbon is the foundation of soil's economy, but complex protein production is the peak of its economic cycle. It is the zenith of soil civilization, as we know it. The greater a soil's carbon reserves are, the more that soil microbiome's protein economy has room to grow. Dense and diverse soil microbiomes expertly build complex proteins as well as break them down into simpler amino acids. If carbon reserves represent a soil microbiome's capacity for work, then amino acids might be thought of as a common, global currency that functions in every life system. The building blocks for all protein and DNA synthesis, amino acids, serve as a primary food source for microbial, plant, and human life. The larger and more complex a soil microbiome's carbon reserves are the more resources it has to build organic proteins into the soil biomass. Like the grain reserves of Egypt, healthy soils also stock up their carbon reserves for future times of need.

Farming, by nature, is an exercise in faith, and control is something farmers, historically, have had to live without. For the thousands of years prior to the Haber-Bosch process, farmers fertilized their fields with whatever resources were immediately available to them. Generally, this meant manure of whatever variety was handy. The healthier the individual it emanated from, the richer in nutrients the manure would be. More an art than science, fertilizing with manure meant applying manure to the plant at just the right time. Manures that are too fresh will burn plant tissues with excessive nitrogen. Manures that are too old have already lost their nutritional value into the atmosphere and environment by evaporation and leaching. No one knew the actual chemical composition of a batch of manure let alone how much of a particular nutrient a fertilizer added to their crops.

Post WWII, mass-produced, inorganic, Haber-Bosch nitrogen began to be paired with inorganic phosphorus and potassium creating the first synthetic, inorganic Nitrogen-Phosphorus-Potassium or NPK fertilizers, and suddenly, for the first time in all of human history, we could control the specific amounts of individual nutrients we applied to crops and easily purchase more, if needed. The name, inorganic fertilizer, true to chemistry nomenclature, simply implied that unlike traditional manures, these fertilizers didn't contain any living or formerly alive ingredients. By definition, inorganic fertilizers were extremely simplistic chemical structures, containing little to no carbon.

Traditionally, agriculturists believed that if a little fertilizer was good, a lot more was better. This hailed from the era when fertilizers generally came from manure that farmers had on hand. With the birth of the Green Revolution though, synthetic, inorganic fertilizers had to be budgeted for and purchased. The question soon became, "How much NPK did plants need and in what ratio?" The yardstick to determine how much synthetic, inorganic NPK we needed to apply was by measuring how much was already in the soil and then calculating how much nitrogen a crop would need above that. When in doubt, it was deemed better to have too much fertilizer than to have a slow-growing crop. Troublingly though, many crops appetite for synthetic, inorganic, NPK fertilizers seemed both voracious and insatiable. Increasing amounts of nitrogen were almost constantly needed to produce the same yields as before. In 1961, about 11 million tons of synthetic nitrogen was produced and sold as fertilizer.[4] Today, that number is closing in on 200 million tons.[5]

Simply because a soil tests as deficient in a nutrient doesn't mean that the plant growing in that soil is also deficient. In one European study, vineyard soil tested as low in available inorganic phosphorus. Logically, we would presume that the grapes grown in these soils were also deficient in phosphorus. However, when tested, the grapes had all the phosphorus they needed. According to our inorganic soil theory, this shouldn't have been possible. How had the grapes had the nutrients they needed if the soil was deficient? Their roots were covered with Mycorrhizae. More amazingly, the Mycorrhizae seemed to intuitively know which nutrients and in what amounts to provide different species of grapevines. Our sophisticated soil tests tell us how much is of a nutrient is in the soil. The Mycorrhizae don't have to test the soil; they know exactly what and how much each plant needs. They will hunt down and retrieve whatever their plant needs. Even in soils where we don't think there are enough available minerals for the plant, the Mycorrhizae can transform undigestible nutrients into digestible forms.

Sometimes popular science supports and promotes a working theory as a fact that is ultimately disproven. In the early 20th century and alongside the Haber-Bosch debut, a single study was conducted to determine whether plants preferred organic (containing carbon) or inorganic (without carbon) nitrogen. The study concluded that plants definitively preferred inorganic nitrogen and that it was naturally abundant. Thus, organic nitrogen in the form of complex proteins and amino acids weren't believed to be of any use to plants. Dramatic yield increases produced by new hybrid crop species following synthetic, inorganic nitrogen applications seemed to confirm this shaky science. Thus, the stage was set for an explosion of inorganic nitrogen usage and broad spread acceptance of the simply named, Nitrogen Theory. Only later on did it come to light that the environment chosen for the study was highly polluted with industrial waste, but not before a flawed Nitrogen Theory had been taught, as fact, to generations.

Now we know that only sick and dying soil microbiomes contain abundant quantities of inorganic nitrogen. In healthy farm soils, even the most basic forms of organic nitrogen are only a short-term by-products of amino acid and subsequent complex protein production. In healthy soils, protein creation cycles are constantly taking place and unincorporated nitrogen typically doesn't stay unincorporated for long. These complex proteins and amino acids, unlike basic nitrogen, are bound by lots and lots of carbon. Thus, complex protein creation, within the soil, requires not only nitrogen, but also abundant carbon resources. Hence, farm soil prefers a 12:1 carbon to nitrogen ratio to keep steadily producing these complex proteins and amino acids and to properly feed its plant partners.

The same essential amino acids that the human body depends upon function as the currency of the soil. Unsurprisingly, when given the choice, plants, like people, also prefer to consume the vast majority of their protein in amino acid form. Free amino acids don't last long in the soil microbiome. They are gobbled up within minutes to hours, and in healthy soils, this process is occurring constantly.

In nature, seasonal plant decomposition and animal manures slowly and steadily add carbon back into soils along with nitrogen. When synthetic, inorganic nitrogen or ammonium is applied to farm soils, soil microbiomes must turn to their own carbon pantries, glomalin and soil acids, to build the complex proteins and amino acids they thrive upon. If soil microbes don't have enough carbon, they can't build complex proteins and amino acids to expand their soil kingdoms[6]. Like a besieged city whose starving citizens are searching for a morsel of flour to make bread with, soil microbes stripped of their carbon savings will hunt down free soil amino acids and begin breaking them apart to salvage and consume the carbon and other nutrients they need to survive.[7] What's left over? Basic, organic nitrogen in the form of ammonium.

In a complete reversal of the theory we've taught and built our agriculture around for the past 70 years, it turns out that plants actually prefer amino acids over basic, organic nitrogen, and they prefer organic nitrogen over inorganic nitrogen.[8] Yet, we've been heaping synthetic, inorganic nitrogen on our fields worldwide now for over 70 years.

What happens when we continue to apply more synthetic, inorganic nitrogen to a besieged and carbon starved soil microbiome? As excess nitrogen accumulates, soil microbes continue to hastily pillage their carbon pantries for any remaining complex protein, amino acids, and fatty acids they can break apart to get carbon from to balance their own diets with. If they run out of enough carbon, they die, and the carbon to nitrogen ratio shifts even further towards nitrogen dominance, intensifying the cycle and creating an increasingly hostile environment towards any remaining soil life.

Natural hunters and gatherers of complex proteins and amino acids, Mycorrhizae don't absorb either synthetic nitrogen or phosphorus well,[9] and are highly sensitive to excesses of both in their environment. At best, as nitrogen accumulates in the soil, Mycorrhizae growth becomes stunted, and they colonize fewer and fewer plant roots. At worst, overzealous or repetitive applications of synthetic, inorganic NPK fertilizers can quickly obliterate any existing Mycorrhizae. Earthworms, tiny nematodes, and insects, the long-distance truckers of the soil that transport sensitive fungal spores respond to excessive ammonium as the poisonous, volatile gas, that it is, and simply die.

Providentially, the soil microbiome has an emergency relief valve to remove excess nitrogen and phosphorus. In a desperate attempt to maintain soil homeostasis, the soil microbiome releases excess nitrogen back to where Carl Haber got it from, our atmosphere. That's where the majority of agricultural synthetic, inorganic nitrogen winds up. Around 60% of nitrogen fertilizers applied to the soil are either vaporized into the atmosphere or washed away into streams and creeks. Only about 40% of NPK fertilizers applied to a field are absorbed by crops[10]. The other 60% ultimately fertilizes our skies and oceans.[11]

As the soil microbiome shifts away from being a healthy, thriving civilization towards nitrogen dominance, plants are no longer protected and fed the balanced diet of complex proteins, fatty acids, minerals, and antibiotics that a soil microbiome inherently provides. Plant survival mandates increasing reliance upon synthetic, inorganic chemicals from a fertilizer sprayer. Like transitioning from a locally grown, freshly harvested, homecooked meal to being tube-fed in a hospital, helpless and unprotected plants must either depend upon synthetic, inorganic fertilizers or die. Tube feeding synthetic nutrients may cause rapid growth in plants, but it doesn't offer the nutrient and mineral density that a healthy soil microbiome can. We have traded short-term, crop yields for long-term soil health.

Soil microbiomes are strong and will battle for their survival. Despite plowing and synthetic, inorganic nitrogen halting the production of natural carbon reserves and devouring existing ones, in many places a remnant of the soil microbiome lives on, albeit diminished in size and strength. No bacteria and no fungus equals no carbon reserves. Glomalin and humic and fulvic acids, the carbon and amino acid storehouses of the soil, are no longer being built and stocked. The mini-aquifers of humic and fulvic acids and Glomalin that store water alternatively cooling plants in summer, insulating them in winter, and staving off the effects of drought are no longer being built. Unsurprisingly, as these carbon reservoirs are consumed, the remaining sand, silt, and clay minerals don't absorb or hold water as they once did. Instead of pulling water into their structures and then acting like little microbial sandbags shielding the soil beneath from rising waters, absentee soils acids and Glomalin mean increased flash floods that wreak havoc on our unprotected soils. The rainfall amounts of modern day, Leptis Magna and other parts of the Middle East don't appear to have changed much over the years, but it is the soil's ability to absorb water that has changed. The creeping process of soil formation shifts downward to an even slower pace.

As stored sources of carbon become scarcer and scarcer, the soil microbiome becomes increasingly cannibalistic and opportunistic, devouring each other like a microbial, Mad Max battlefield until remaining carbon sources are utterly consumed from the soil. Farm fields lose both the ability to transport and store nutrients. Vital nutrients, such as calcium and potassium, slowly ooze out of the soil and are washed away.[12] No longer shielded by humic and fulvic acids and unanchored by matted strands of Mycorrhizae, remaining soil minerals, bereft of both their inhabitants and defenders, readily erode away when confronted with abrasive elements such as wind and rain. The former dwelling places and mighty microbial ruins are easily ransacked and swifted away by wind and water. Like the abandoned ruins of Leptis Magna, sand, silt, or clay, soil's mineral attributes, may still be used to physically identify the old haunts, but the glory, the industry, the growth, and the life are all gone. Our once healthy, vibrant soil microbiomes are being supplanted by tube feeding with synthetic, inorganic nitrogen.

Any farmer today would have to be over 100 years old to have known farming and soils before NPK. For farmers who grew up after 1950 and the introduction of synthetic, inorganic fertilizers, the deadly effects of these now every day tools of the trade are baffling. How could the Green Revolution's high-yielding farming techniques that saved millions of lives be simultaneously destroying our soil?

Healthy, fertile soils largely haven't existed in 70 years now. Most of our benchmarks for judging soil health are askew. Today's farmers have never felt them, never smelled them, and never tasted them. When sick, depleted soils are all you've ever known, recognizing that soil as sick and depleted can be challenging. Overzealous plowing and synthetic nitrogen applications have brought soil microbial death on a massive scale. For decades, we saw the aftermath of the battle drifting down the mighty Mississippi without understanding what was happening because we didn't know the soils had been alive. As unique and ancient soil microbiomes died off from field to field, we only saw that plants were needing more and more nitrogen to produce the same yields. We couldn't see that every inorganic nitrogen application only further depleted final carbon reservoirs. The result was unbridled erosion. Erosion is not the malady that kills topsoil; it is the hearse that carries off the deceased remains following carbon depletion. This is not an American phenomenon. It is an impending global disaster. Farm soils devoid of a healthy, balanced soil microbiome are farm fields on borrowed time.

Erosion isn't a cause. It's a result. Erosion didn't cause the loss of 70% or more of one of our most precious resources. It's the result of soil microbial death. It's the dispersal of the mineral estate that microbial life once held together. Erosion is merely an undertaker employing wind and water as the hearse. It's as if we've spent decades, even millenia, studying the redistribution of personal items following an estate sale asking how personal effects will be packed, where they will go, and how they will get there. We've contemplated these effects all the while overlooking the singular event that brought us here in the first place, death.

Chapter 4 Fast Kill, Slow Kill

"Dip the apple in the brew. Let the sleeping death seep through. Look! On the skin, the symbol of what lies within. Now, turn red to tempt Snow White to make her hunger for a bite."

~Disney's Snow White, 1937

Known as the poison of kings and the king of poisons, arsenic developed its reputation as a discreet homicidal agent before the existence of the Roman empire. Tasteless, odorless, and undistinguishable from food poisoning, arsenic was entirely undetectable by coroners prior to scientific advancements in the mid-1800's. No longer employable in the spheres of revenge and inheritance, arsenic took its skillset to a new territory, American farm soils, and locked in on its new target, the Colorado potato beetle.

Early successes as an insecticide fostered increased demand for its services and additional crop insect targets were added to the roster.

Chemical companies branded and differentiated their arsenic products with cheery names such as London Purple, Scheele's Green, and Paris Green, reflecting their product's color. To increase potency, lead was often added to the mixture along with sweet molasses and bran mash to entice insects to eat it. Apply liberally, farmers were told, and they did. Sometimes by truck, but often with their bare hands.

Unfortunately, insecticides aren't species specific. They are dosage specific, meaning their lethality isn't restricted to insects. Tiny insect bodies are more quickly overwhelmed by smaller quantities of insecticide than larger, living creatures. Nonetheless, insects that do survive insecticide spraying have an advantage we don't, their generational length and prolificacy. Generation can beget generation in anywhere from a few days to a few months. What's more, they are highly prolific. Depending on the species, female insects can lay from a couple hundred to tens of thousands of eggs per day. Over a lifetime, some female insects will lay millions of eggs. Of these eggs, some will survive pesticide applications. Some will be just far enough away from a sprayer, perhaps resting on a neighboring fence post or in a culvert, to avoid a lethal insecticide dose. Other individual insects may survive through a slightly superior, genetic ability to detox the insecticide from their systems. These few surviving insects rapidly bequeath a swarm of progeny with now, relatively concentrated, superior detoxification genetics, and each successive, surviving generation becomes better and better at flushing that particular insecticide out of their bodies. Thus, unlike humans, who have exponentially longer generation times and a relatively, few offspring, insect species can quickly adapt to new insecticides. We call these genetic adaptations, resistance.

In the mid to late 19th century, as cultivated farmland increasingly emerged from former American forests and frontiers, arsenic applications to crops increased and naturally, so did resistance. A vicious cycle ensued, and ever-increasing quantities of arsenic had to be applied to prevent crop decimation. While insects were building resistance to the increasing concentrations of arsenic though, human beings were not.

A tiny bit of arsenic won't kill you. Small amounts of natural forms of arsenic are even present in drinking water. Arsenic is dosage specific after all. Decades prior to the release of Disney's *Snow White* in 1937, American apple orchard arsenic concentrations were so high that Americans were reportedly, literally, keeling over from even a single, bite of an American apple. By the 1920's, the U.S. government began quietly pulling arsenic-saturated apples from the market and then destroying them.

As the horse and plow gave way to the tractor and arsenic began to accumulate in American farm soils, came the 1930's Dust Bowl. Great Plains residents stuffed and restuffed blankets, sheets, and towels around window and door frames in attempts to block out encroaching dirt. The dirt got in anyways. Grasshopper hoards estimated at 23,000 per acre swarmed in and descended upon the scant remaining crops that humans and livestock alike looked to for survival. Mankind's survival seemed to hinge upon insect eradication. Despite the now well-known risks, we felt we had no choice and turned once more to applying more arsenic.[1]

How liberally did early 20th century farmers apply arsenic pesticides? Pesticide sales and applications have traditionally been both sporadically reported and under reported and are even today. Records from the early 1900's through the 1940's indicate that at least 30-90 million pounds of arsenic-based pesticides were sold annually for application on American soils. Conservatively, around

1,200,000,000 pounds of arsenic blanketed the 400 million acres of American cropland that spanned from New York to California over the first 4 decades of the 20th century. Three pounds per acre over 40 years may not sound significant, but arsenic doesn't break down in the soil. It doesn't wash away. It's persistent. It began to bioaccumulate. A new solution to the insect problem had to be found.

In the Fall of 1943, war refugees were hitch-hiking their way to Naples, Italy where U.S and Ally troops had just fought and won a tenacious and costly battle to free the city from German powers. Once arrived, the refugees, along with Naple's natives, sought safety from overhead bombing through underground living. Natives and refugees alike crammed into Naples' massive, underground, cave systems. The caves naturally lacked sanitation and were overcrowded. In the midst of squalid living conditions and with the people bordering on starvation, it quickly became apparent that on the heads of refugees had come another less welcome guest, typhus carrying lice.

By December, nearly 400 Naples residents had contracted typhus. To protect U.S. troops that were guarding the city, the U.S. Army naturally wanted to quell the growing epidemic. The Rockefeller Foundation and along with a company called Monsanto offered to help create a solution. They had a new powder that might be a miracle cure for lice. Monsanto's previous successes producing Aspirin and food additives inspired confidence in their brand. Armed with this newly engineered, lice-killing powder, the U.S. Army set up forty, massive, walk-through spraying stations around Naples that were designed to dust up to 100,000 people a day. With the Rockefeller's as chief financiers, by February of 1944 over 3.2 million people had been dusted with the chemical now known as DDT. Typhus and lice vanished. The success was unprecedented. The epidemic was declared over, and the operation ended.[2]

A little more than a year later, like synthetic, inorganic nitrogen, Monsanto's DDT emerged from WWII in need of a job. Given its demonstrated success in Naples, its resume was quickly seized upon to eliminate the American agricultural insects that arsenic and lead had failed to conquer. If arsenic was applied liberally, DDT was applied lavishly. Frequently, DDT was mixed with diesel oil or kerosene in a 1 to 9 parts mixture. In one formulation, it was mixed with paint to apply onto farm buildings. Whether in the wilderness, on the farm, or within the home, it was regarded as not only effective but harmless to adults and children, and like arsenic, farmers often handled it with their bare hands. As air power became more accessible, we began spraying it down from crop dusters.

The timing must have seemed conspicuous. In the early 1950's, just as agronomists were trying to make sense of the massive, Mid-Western, farm soil exodus, they appeared on the horizon. Like Hannibal and his troops, foreign, invasive, and previously unknown insect species rose up in mass to encircle, devour, and decimate crops. Worms, beetles, and moths, whether accidentally imported by cargo ships or planes across oceans or swept in by weather systems, began to hungrily devour crop roots, stems, leaves, and produce. No longer hindered by the natural predators of their native regions, these foreign worms, beetles, and moths began to hungrily devour food crops. Along with them came previously unknown strains of blight and mildew wreaking similar havoc. *"Diseases of progress"* were how one chemical company labeled them.

Cashing in on the heartland's fears, Hollywood both amplified and made light of them by producing insect invasion flicks. Following King Kong and preceding Jurassic Park, in true apocalyptic format, these films frequently paired a wise scientist with a love interest racing to stop these seemingly unstoppable insects before the inevitable decimation of the earth's crops and entire human population. *"The*

Beginning of the End!", "Them!", "The Deadly Mantis", and *"Insect Fear"* to name a few, left movie goers with the petrifying, albeit illogical, message *"Kill one and two take its place!"*

> *"Books and magazines spoke of the insidious efficiency of insects, relentlessly consuming our crops, infiltrating our homes, and undermining our health; the arthropods, one entomologist direly warned, would 'inherit the earth unless man abandons war and turns his martial interests to killing pests.'.... Commentators regularly wrote of the 'war on insects,' the invasions of pests, and the battles to exterminate unwanted species. .... 'Would it not be funny,' one U.S. representative observed, 'to spend all these billions of dollars fighting communism and building the atomic bomb and then be eaten by the Argentine ant?'" ~William Tsutsui[3]*

The Green Revolution's phenomenal yields meant immediate security and prosperity, but only if they weren't wiped out by simultaneously increasing insect damage. The mantra remained that human survival and a rapidly expanding global population depended upon insect obliteration.

The name DDT is somewhat of a misnomer. Its deadly active ingredient is actually chlorine, the same as used in drinking water and swimming pools. That's probably why farmers felt comfortable amply applying it to their crops. How different was applying DDT to fields from municipalities applying it to drinking water? Chlorine sounds a lot like chloride, and in chemical terms, there's only one electron difference between them. The physical consequences of this single electron though are vast. Chloride occurs abundantly both in the natural world around us and is essential within our own human bodies, but chlorine is different. Chlorine is only naturally found in the deep recesses of the earth's crust as a highly potent, poisonous gas.

There Might Be Hope

By the chemical world's standards, DDT isn't considered to be particularly toxic. Each and every DDT molecule contains only 5 chlorine atoms, hence the name Di-*chloro*-phenyl-tri-*chloro*-ethane. Aspirin is more immediately toxic. The primary problem with DDT and other forms of chlorine, for that matter, is generally not your initial exposure. The problem is your residual exposure. Pound for pound, the body is better at ridding itself of small doses of Aspirin than the chlorines found in pesticides like DDT and other PCB's or Polychlorinated Biphenyls. Chlorine is part of a group of chemicals known as Halogens that includes Fluorine, Bromine, and Iodine. Because Iodine and Chloride are essential to both animals and some plants, multitudes of receptors for them exist within animal and plant tissues. Unfortunately, due to their structural similarities, both animal and plant bodies will mistake poisonous halogens for the essential ones, and once absorbed, they become extremely difficult to eliminate and may build-up over time in an individual. It is from this family of chemicals that we often hear the term "Forever Chemicals."

In 1972, after almost thirty years, an estimated 1,350,000,000 pounds of DDT had been sprayed on a mere 900,000,000 acres of American farm soils and foods. Public perception of DDT soured in the wake of Rachel Carson's book *"Silent Spring."* Published in 1962, the book called public attention to the noticeable decline in songbirds which had fed on insects poisoned with DDT. Public reaction was so strong that the EPA banned DDT in the U.S.[4] Largely an appeasement to public perception, the ban removed DDT specifically and other PCB's, but it didn't stop the U.S. and other countries from using halogen-based pesticides. Names changed. Formulations changed, but pesticides actually grew much more, not less potent. They had to. We weren't winning the war on insects. Just as with arsenic and lead, insects were developing resistance to halogen-based pesticides.

With billions of pounds of arsenic, lead, and halogens having been poured out onto our American farmlands without much result, alternative methods became essential. Monsanto, once again, came to the rescue with a novel solution. Bacillus anthracis, the bacteria responsible for anthrax and Bacillus cereus, a bacteria responsible for food poisoning, eye infections, and periodontal disease are genetically incredibly alike. They have another close relation known as Bacillus thuringiensis or Bt. These genetic similarities between all three are so striking that some scientists have suggested that all three are actually the same species. Named after Thuringia, Germany, the location of its discovery, Bt first caught scientist's attention when they observed its rapid, lethal effect on its first, observed consumer, a moth larva.[5] Lightning struck. Why not replace bioaccumulating, chemical pesticides with a bacteria to kill insects? What could be harmful to the environment about that? But how would a bacteria be sprayed on fields? It wouldn't. Bacillus thuringiensis would be bred into food.

European farmers have a variety of names for it. In France, it's *broussin* meaning burl. In Italy, it's *rogna* for mange. German farmers curse it with a virtual thesaurus of names including *ausschlag* (rash), *mauche* (ugly), *krop* (goiter), *raude* (mange), and perhaps most accurately "*krebs*" meaning cancer.[6] Officially, it's known as crown gall, and it appears as a cluster of regular plant cells grown wild into grotesque, cauliflowered bulging lumps. Plants, like people, occasionally get tumors, big ones. A pathogenic bacteria causes them. At first, we thought this cancer-inducing bacteria only infected woody stemmed plants, but scientists quickly discovered they could transfer this bacteria to a variety of softer, herbaceous plants as well. Then, we learned that these pathogenic bacteria had a truly remarkable ability to enter a host plant and transfer genetic material from the bacteria to the plant. This bacterium is what causes the tumors.

The ability to transfer genetic information from one species to another was unknown prior to this. Animal bodies, in fact, are designed with a plethora of protections to resist invasion from foreign species genetics. Even when genetic transfer between similar species does happen, it doesn't progress to multiple generations. Wild herds of mules or packs of ligers don't exist, and with good reason. They are naturally sterile. Very rarely, the offspring of two different species, such as a mule, has produced offspring with one of the parent species, but that is where it stops. There are no further offspring. Scientists haven't failed in this pursuit for lack of trying. Even if scientists are able to create embryo offspring from a hybrid and non-hybrid species and providing the female body does not reject the foreign DNA immediately, the female body of each species is so unique that the female will not be able to carry the offspring to term. Each mammalian species has a unique umbilical system or placental cotyledons. Both the number of cotyledons and the shape are unique to each species. Like electrical outlets and plugs on the same continent versus those on a different continent, the cotyledons of a mother and baby from the same species match, but a mother and baby from different species won't.

Plants also have multiple defense systems to protect themselves and resist the transfer of genetic material from outside species, regardless of the origin. These environmental safeguards protect against environmental chaos and destruction. Changes to species DNA in nature are typically negative and harmful. This was true in this case also; the genetic change caused cancerous tumors on the plant.

So, how did this incredible DNA transfer happen in spite of these protective mechanisms? Some bacteria and occasionally plant cells were discovered to have a small disk or mini-genetic portal in their cells, known as a plasmid, which could receive genetic material like a book drop or a thumb drive. In nature, the genetic material these plasmids introduce would be nearly always harmful to the plant by creating cancer. Thus, the smaller the plasmid, the more protected the plant was. This particular bacterium has large plasmids though. Scientists found that they could utilize these large plasmids to transfer the genetic material they wanted into the plant's cell. Perhaps, they could have named it Transbacterium, to reflect the bacterium's primary function, transferring genetic material from one species to another. Instead, they named this bacterium Agrobacterium, a bacterium that would transform all of agriculture.

Agrobacterium was first put to work by transferring Bacillus thuringensis into tomatoes. Tobacco, corn, potatoes, soybeans, cotton, and rice soon followed. When compared with continuing to saturate our soils with arsenic, lead, and halogen-based pesticides, Bt crops seemed like the lessor of two evils.

Unarguably, in the early 1980's, genetics were conceived to be much simpler than we now know them to be some 40 years later, and even some 20 years post-introduction of commercial Bt crops, it was widely acknowledged that the genetics of Bt crop engineering were poorly understood. Though genetic engineers had access to this amazing genetic portal, they had little control in the insertion of these genes, meaning that inserted genes often flipped, were broken, or mutated in the process. Such insertions also often inadvertently were observed to change the expression levels of natural genes.

"In MON 810 corn, for example, the promoter sequence (that turns on the foreign gene) was only partially incorporated into host DNA.
The transgene in MON 863 corn underwent some degree of mutation as well, and Syngenta's Bt-176 corn was supposed to create the Cry1Ab form of Bt, but.... analysis 'carried out both by French and Belgian government scientists' showed that the transgene had

only a 65% similarity to the intended protein." ~ "Genetic Roulette", Jeffery M. Smith,

Nevertheless, Bt crops were touted as a miracle to reduce dependence upon chemical insecticides. For a time, that was likely true. While they never replaced insecticides entirely, they did lessen our reliance upon them, but as with arsenic, DDT, and other halogen-based pesticides, Bt crops turned out to be just another 30-40 year stop gap. Bt resistance was always inevitable. Everyone knew that. This relentless chemical and genetic assault on insects, has albeit unintentionally, continuously and selectively bred insects with enhanced detoxification abilities. In the time it takes a pesticide company to engineer, test, approve, and market a new formulation, generations of insects have matured and reproduced young entirely immune to existing solutions. Some species have emerged with complete resistant to a new insecticide within as little as 2 years.[7] At least 580 insect species are currently resistant to at least one insecticide. Some insect species have developed resistance to more than 90 different insecticides. Scientists can't keep up.

Despite the miracle of Bt crops, the Halogen family of Chlorine, his beefier brother, Fluoride, and their cousin Bromide still dominate a broad array of widely used insecticidal agents today. Although insecticides are classified primarily by their mode of killing, scientists have, even lately, candidly admitted that they aren't completely certain how these chemicals work. Even after nearly a century of usage, the mechanisms of insecticides and their cocktails are still being theorized and debated. One thing is certain. We are losing this war, and we are losing it so badly that arsenic is being considered for reintroduction on American farm fields. We've come full circle. Famine by insect plague is no less an impending reality today than it was when we began pouring halogens on fields in the mid-20th century. If anything, famine by insect plague is more imminent today than it was 70 years ago.

Yet for the billions of pounds of arsenic, halogens, and Bt genetic material that have saturated our only 900 millionish acres of American farmland over the past 100 plus years, insecticides actually comprise a small minority of pesticides used. Of the 700 or so agricultural pesticides currently utilized in American agriculture, the majority are herbicides. They kill weeds.

Chapter 5 An Unloved Flower

For Dust Bowl survivors, plowing came to equal erosion. Erosion came to equal dust storms, and dust storms meant hunger and scarcity. Just like Adam and Job though, they plowed for a reason. So do we. Among other reasons, plowing eliminated unwanted plants from competing for sunlight, nutrients, water, and space with the food they wanted to grow. We call those unwanted plants weeds. Plowing rips apart any existing plant root systems and buries them alive under the soil, giving the young seedlings we plant a head start. To feed the world's rapidly growing population, weeds had to be plowed under, but to save our soils, we had to find a way to farm without plowing.

Solving this age-old agricultural problem wasn't how the world's best-known herbicide came into being though. In 1942, in an effort to preemptively answer the legitimate threat of biological warfare, the U.S. Secretary of War selected George Merck to direct research on technology that might be utilized in biological warfare. In modern echoes of the Romans salting Carthage, one promising theory was to destroy Axis food crops by spraying them with herbicides from planes. Among nearly 1,100 chemicals being experimented with on crops at Camp Detrick in Maryland, two drew considerable interest. Recently synthesized 2,4-D and its cousin 2, 4, 5-T showed promise. According to Merck in a 1946 letter, it was *"only the rapid ending of* the war that prevented field trials, in an active theater of synthetic agents that would, without injury to human or animal life, affect the growing crops and make them useless."[1]

In 1945, 2,4-D made its entrance into commercial agriculture, and as agricultural historian, Gale Petersen, notes that *"the public's enthusiasm for the newly, unveiled herbicide was equaled by their universal ignorance on the chemical and how it worked."* Such an enthusiastic reception prompted more chemical companies to begin producing their own 2, 4-D herbicide formulations. In only 2 short years, 30 unique 2,4-D based herbicides were available and over 5,000,000 lbs. had been sold to American farmers. Since the 1950's, about 30 million pounds of 2,4-D have been steadily applied to American agricultural crops every year.[2] Because pesticide sales have historically been both loosely reported and underreported, the figures are certainly on the low side. By the mid 1960's, production had grown to 10 times that and 2, 4-D was found in 6,000 different formulations. Said by some *"to have initiated the agriculture revolution,"*[3] 2, 4-D's success at obliterating weeds in pursuit of untainted crops ensured its rapid ascent to becoming one of the most widely used pesticides in the world.

"Weeds decrease the yield and lower the quality of crops in gardens, on farms, on the plantation, and on the range. Weeds disfigure the land, affect the national health, and the national pocketbook. Death to weeds!" ~ *Dow Chemical Company, "Death to Weeds", 1947*

George Merck didn't live to see the mobilization of the work he began with 2,4-D in WWII, but 2, 4-D and 2, 4, 5-T would ultimately both realize their original purposes. In the end, they were sent not to annihilate field crops in the European theater, but to defoliate the jungles of Vietnam. Roughly 20 million gallons of Rainbow Agents accomplished their mission, eliminating enemy cover and clearing the way for U.S. base camps. "*Developed for war, but designed for agriculture*"[4], the Rainbow Agents came in 7 different formulas. Agent Blue was based on arsenic, but Agents Green, Purple, Pink, White, Orange, and Super Orange, were all based on either forms of 2, 4-D and 2, 4, 5-T singularly or in combination.

Like DDT, the "D" in 2,4-D stands for "Di-*chloro*" or 2 chlorine atoms, its primary lethal agent. 2, 4, 5-T's or 2,4,5-Tri-*chloro*-phenoxyacetic acid also relies on chlorine as its lethal agent. The mechanisms by which chlorine causes death differs between herbicides and insecticides, but the result is the same. Often, the particular potency of these formulas can be attributed in part to their structural similarity to essential plant growth hormones. These herbicidal Trojan horses fool the plant into welcoming them in as essential hormones, and that is precisely what they do.

As with the insecticides though, from 2, 4-D's introduction in 1945 through the 1970's, weed resistance grew. So, we began mixing herbicide cocktails. Layering different formulations of chlorine-based pesticides together increased their individual efficacy. By 1970, over 100 different herbicides were in general use.[5] As with the insecticides, most were halogen-based thus compounding our bio-acculumation problem. A solution to growing weed resistance and poisonous halogen accumulation in the environment had to be found. Once more, Monsanto led the way. In 1974, they brought a novel idea to the table, simply called glyphosate. Commercially, it would most frequently be dubbed Round-up.

The unnatural union of two naturally abundant and essential nutrients, glyphosate, is disarmingly simple. It is the combination of glycine and phosphate, both essential nutrients to plants, animals, and people. Glycine is a neurotransmitter, serotonin producer, and essential DNA building block. It supports mammalian mental health, sleep, memory, and the digestion of fatty acids. It also regulates the body's immune system and heals inflammation. Phosphate is also essential. Bones, teeth, and cellular membranes all rely heavily upon phosphate, and without it, muscles won't contract. Nerves won't function. How do two such brilliant and vital ingredients combine to form a deadly herbicide?

In plants, one of phosphate's jobs is to build a multitude of amino acids including the essential amino acids tryptophan and phenylalanine. This amino acid building process is known as the Shikimate Pathway, and it's glyphosate's job to stop it. The Shikimate Pathway is tricked into accepting glyphosate as simple phosphate, but with the addition of glycine, it's much bigger. Like an assembly line where one slightly wrong part clogs the entire line, glyphosate settles into phosphate's normal space and halts any further production line progress. Essential amino acids tryptophan and phenylalanine can't be made, and the plant starves to death for lack of these essential amino acids. The enzyme that would have been produced by the Shikimate Pathway is called EPSP, and it is essential for the synthesis of aromatic, amino acids and *"almost all other aromatic compounds in algae, bacteria, and fungi."* [6][7] Glyphosate itself isn't lethal because it's poisonous; it's lethal because it creates starvation through nutrient deficiency.

Glyphosate is considered to be the active ingredient in Roundup and is labeled as such, but in recent years, Roundup and other herbicides inactive or inert ingredients have been receiving increasing attention. It is actually these "inactive ingredients" that comprise the bulk of herbicides today, and because they are considered to be "inactive," companies do not have to disclose these ingredients. These inactive ingredients mix and marry with the active pesticide and water to help them slip smoothly through a sprayer and stick to the plant's foliage. They are also designed to enhance absorption and persist upon the leaves of the plant. [8] Because they have been considered to be both benign and proprietary, chemical companies have not been required to divulge these ingredients. Until recently, no one considered testing these ingredients for safety. New research though is showing that these "inactive" ingredients are often far more lethal than the actual active toxin. Heavy metals such as arsenic, lead, nickel and petroleum products often comprise the bulk of these inactive carrier agents. One such petroleum product, POEA, is considered to be up to 2,000 times more toxic than Glyphosate itself.[9] Farm fields and farm soils are literally being coated, over and over and over again, with variations of petroleum and heavy metals.

 Despite all of this, because a new generation of plants sprouts up every year, glyphosate resistance, like insect resistance, was inevitable. Knowing this, scientists wasted no time in beginning to develop the next solution, and as with insecticide halogens, the immediate answer was to apply more glyphosate. From 1945 until 1996, the amount of herbicides that could be applied to a field were dictated by the crop's life cycle. Herbicides generally had to be applied before planting or after harvest. Spraying during the growing season might kill crops. With Agrobacterium in their pockets though, scientists needed only to reach through the genetic portal to engineer a crop that could survive without tryptophan and phenylalanine. They could do what had never been done before, spray growing crops with herbicides. The first genetically altered crops that could survive without the Shikimate pathway were named Round-Up Ready crops. Later, as patents for similar crops proliferated, these genetically modified seeds would become generically known as GMO's. For the first time in human history, seeds were patented.

 If the wind carried seeds or pollen with patented seed DNA from a farmer's field to his neighbor's field and the plant grew, farmers could and were often sued for theft by the seed patent owners. In fact, some 410 farmers and 56 small businesses across 27 states were sued by Monsanto for patent infringement from *"genetic drift"*[10,11] or reusing seed from a previous crop, which was forbidden. Many farmers lost the lawsuit and then their farms.

In the 1980s, we barely understood Mycorrhizae and the soil microbiome better than Roman farmers, but soil compaction had already been connected with diminishing yields. Reasonable and rational fears of a second dust bowl loomed large as the best of America's farmland poured into the Mississippi and emptied into the Gulf of Mexico. The answer to the plow, the answer to Leptis Magna, the answer to the Dust Bowl, the answer to our Great Plains washing down the Mississippi—it was all wrapped up in glyphosate and GMO's. They had to work. Our future depended upon it. We couldn't keep plowing like we had been. We knew we couldn't keep pouring halogens and arsenic into our farm soils.

By genetically engineering crops to withstand the toxicities of pesticides, farmers wouldn't have to till anymore to eliminate the weeds that fought their crops for limited sunlight, water, and nutrients. The method was simply titled "no-till" farming. Although human history repeats itself frequently, the title of being the first generation in all of human history to bravely attempt no-till farming on a commercial level surely belongs to us.

Fields planted with Round-Up Ready crops could not only be sprayed with glyphosate before and after the growing season as they had been in two previous decades. They now could also be sprayed two to four times during the growing season without killing the GM crop. Predictably, sales went through the roof, and to a point, farmers increasingly adopted it. Why wouldn't they? No-till farming offers tremendous savings on fuel and labor costs. According to U.S. government estimates, a farmer working a 1,000 acre field can save around $10,000 in fuel and 70 hrs. of work a year by switching to no-till. That's a nice vacation to Bora Bora in the dead of a frozen,

Iowa winter. So, why are only around 20% of U.S. farmers using no-till 40ish years later? Perhaps, many farmers have realized that conventional "no-till" came with fine print.

Some called them Super Weeds. Others called them Franken weeds. Ragweed, Milk Thistle, Cat's Claw Sensitive Brier, Lambs Quarter---not only are these plants not deterred by herbicides; they seem to thrive in spite of them. Two, four, or even six times in a single season, glyphosate and other agri-cocktails rained down upon our farm soils. The more we sprayed, the more weeds grew. Impervious to chemical assault and seemingly wed to the embattled and dying soil microbiome, these unexpected nuisances could not be deterred. Eventual resistance had been anticipated, but not with this speed and tenacity. No one plants them in the middle of crops, yet they emerge and thrive. Between resistance to insecticides and resistance to herbicides, we are at a dead end.... or are we?

Beneath the crop dusters, beneath the tractors, beneath the sprayers, soaked in insecticides, herbicides, and inert ingredients lies what remains of our embattled soil microbiomes. It is here that super weeds love to make their homes. These "nuisances" to our preferred, food crops love poor and dying soils, and they are here to rescue them. Super Weeds, indeed. These super weeds tend to be plants that naturally pull a lot of nitrates, a form of nitrogen, into their leaves. By doing so, they help to rebalance and restore a healthy carbon to nitrogen ratio in farm soils, and thereby begin to prepare the soil for soil microbiome restoration. When they die, their stalks, stems, and leaves will provide food to help add even more carbon to the soil for microbes to feed upon. Some of these weeds also pull heavy metals out of the soil. There is an undeniable beauty to these weeds. In spite of our determination to annihilate them, they are here to save us. They are here to fight for us. They are here to begin detoxifying our soils.

If the Super Weeds aren't actually nuisances, but heaven sent heros, are the insects really as bad as we have made them out to be? There is a group of agricultural scientists who believe that *"all plant and parasite interactions are nutritional in nature."*[12] This small contingent espouses that insect invasions are not bizarre, inexplicable, paranormal heralds of apocalyptic devastation to be battled with all our resources. Insect invasions, prior to the 20th century, generally followed periods of intense cultivation and extended drought. With hindsight, we can see that crop insect invasions served a larger, beautiful, and, ultimately, essential purpose. Like vultures and sharks, these insect invasions were the undertakers for nutrient starved and dying plants. We weren't aware of how unhealthy the plants were, but the insects knew. The theory of trophobiosis espouses that insects only attack plants *"when the plant's biochemical state meets the nutritional needs of the parasite in question"*[13] In other words, insects can tell, intuitively, when a plant is strong and healthy. These plants are unattractive to potential predators, but when a plant is nutritionally weakened, insects do what they were designed to do, consume the plant and recycle nutrients back into the soil microbiome to help ensure its health for future plants. When these insects die, their own bodies will further help rebuild the soil microbiome as its on its path to complex nutritional recovery.

The presence of insects, in mass, ought to be a warning alarm for us that our soil microbiome is failing and our plants are starving for nutrition. Our trouble is not that insects are doing their job in trying to rebalance a failing microbiome; our trouble is that our crops are weakened and sick from generations of living on i.v. drip nutrition.

By means of their inert or inactive ingredients, pesticides, whether insecticides or herbicides are designed to stick to plant's leaves. Plants, like soils, breathe, and inevitably, some pesticide is inhaled into the plant. Some may also be absorbed into the soil and plant root systems.

Whether through leaf, stem, or root, plants absorb pesticides. Their response? First, shock then, like any trauma victim, demonstrable panic. All measurable growth and protein production halts as plants shift into red-alert, detoxification mode, trying to eliminate the poison. Plant cells begin breaking down the complex proteins and carbohydrates of their own tissues back into simpler amino acids, nitrogen, and sugars. Sugars begin pooling in the extremities and tissues begin to swell within the circulatory system. In humans, we'd call this edema, and as with people, it's primarily found in plant extremities or the tips of their leaves.[14] This pesticide-induced edema is not visible to the naked eye, but it's scientifically measurable. Simply because the plant didn't die or wilt does not mean it is unaffected. It just means that the quantity of toxin it ingested was insufficient to kill it. We've long held the absence of plant death to be indicative of plant health, but have we set the bar too low?

This systemic shock from pesticide trauma temporarily increases the amount of simple sugars in the plant's cells making them simply irresistible for insects. Like the aroma of fresh bread or pastries wafting from a French bakery, plant stress beckons insects with the readiness of their favorite meal. We might as well be ringing the dinner bell with every pesticidal pass down the farm row.

Insects are not drawn to a certain plant simply because it is a particular species. Insects don't have a favorite vegetable per se. They are drawn to particular crops and particular fields because the crops in these fields are growing in poisoned, sick, and dying microbiomes. These soil microbiomes lack the Mycorrhizae, humic and fulvic acids, the bacteria and protozoa, as well as the complex proteins, nutrients, carbohydrates, and available minerals that would have grown strong, healthy, and nutrient dense plants that would be less appealing to insects.

In observable science, every reaction follows an action. It may have taken us a while to locate and measure the initial cause, but we are here now. Perhaps halogen gases, petroleum carriers, heavy metals, and genetic manipulation aren't necessary to prevent insects from devouring our crops. We have to target what's going wrong in the plants and the health of our soil microbiome that's making them susceptible and attracting deadly insects. To correct key nutrient deficiencies in the crops themselves, we have to diagnose and correct imbalances within the soil microbiome.

Chapter 6 The Diversified Farmer

"Now Able was a keeper of the flocks, but Cain was a tiller of the ground." ~Genesis 2:2, NKJV

The origins of farming and ranching are as old as Cain and Abel. Farmers, traditionally, stayed rooted on a single parcel of land using a few animals for their strength and manure to grow crops. Ranchers, herdsmen, or shepherds moved seasonally from one grazing area to another to allow their livestock to convert natural forage into meat and fiber. Capable farmers thoughtfully applied their manures back as fertilizers. Wise ranchers kept their pastures from being overgrazed. A natural cycle and rhythm to life existed within the bounds of each endeavor.

Following the Civil War, Northern states and even England were hungry for beef and had the money to pay for it. Texas had an abundance of feral cattle and grass. Emerging railways offered entrepreneurial opportunity for those willing to risk their lives driving the cattle, on horseback, from Texas to Kansas. A roughly 3 month trip, trails were fraught with danger, but the lure was being paid immediately upon depositing the cattle, in cash. Cowboys earned a living. Chuckwagon cooks got ahead, and trail bosses got rich. Along the way north, the cattle grazed steadily as they ambled along. Once arrived at a railhead, they would be shipped by railcar to the Chicago Union Stockyards.

If Texas in the 19th century was a sea of grass, Chicago was moated by an abundance of corn. Once at the stockyards, these feral cattle that had just walked from Texas to Kansas would spend their last 100 days in dusty, wooden holding pens being fattened on corn at Packingtown on the edge of the Chicago stockyards. Not only were butchers able to increase their profit by adding pounds to their final product, but the corn gave the beef a comparatively mild and unique flavor to that of grass-fed beef, which had been the previous standard.

Prior to the advent of the Chicago Stockyards, it's important to remember that butchering a steer was often intensely personal. Families and small landowners either butchered animals they had raised themselves or personally delivered them to a local butcher. Calves were likely to have been cared for and lovingly watched over. Children often played with and doted on them. Not only personal, butchering a steer was also an intensely physical, all-day task, easily requiring 8-10 hours to complete. After slitting the throat, a butcher needed to hoist the roughly, thousand-pound carcass into the air, and then carefully remove its swollen and bacterially-laden intestinal tract from the body without puncturing it and contaminating the meat. Next, a butcher leveraged his body weight, using his bare knuckles to alternatively pull and "fist" or push the hide down to separate it from the animal's body. Finally, he must saw through bone, muscle, fat, and tendon, tediously breaking down the carcass piece by piece.

There Might Be Hope

There was nothing personal about the Chicago Slaughterhouse. By leveraging the power of the assembly line, they could process an entire steer in 35 minutes. A model of efficiency, they moved up to 2,500 steers a day through their facilities. Corn-fed beef that had originated on the wild, Texas plains would then be shipped in refrigerated rail cars to restaurants and groceries on the increasingly, urban Eastern seaboard. From 1890-1930, Chicago supplied up to 80% of America's beef.[1]

The invention of barbed wire gradually diminished the open range and herds of wild cattle. The cattle trails were replaced with more railroads, and then, the highway replaced the railway. The Chicago Union Stockyards and Packingtown shut down, but the slaughterhouse assembly line model was here to stay. America had fallen in love with corn-fed beef. So, the slaughterhouses followed the corn and migrated to the Midwestern Grain Belt to save money on shipping feed. Further and further, technology and mechanization separated American citizens from what once had been an intimate connection with their food.

It wasn't just the beef industry that saw dramatic changes during the 20th century. Lamb, which once had been a staple of American diets, became unfashionable following WWI. Returning GIs had eaten their fill of cold, greasy, canned English mutton war rations. Meanwhile, a pig and some chickens had once been backyard American staples for every household. As Americans moved to the suburbs, they left behind their pigs and chickens. Swine and poultry farms became increasingly consolidated and themselves moved from outdoors to indoors within massive, climate-controlled buildings on concrete and plastic floors. Laboratory-like and fastidious, these houses maximized sanitation and minimized disease. They also cut swine and poultry off from a diet that had for millennia, held an intimate reliance upon the soil. In fact, baby piglets that once naturally received essential iron by snarfling in the soil had to start receiving iron injections once they began to birthed on concrete.

Last of all, were the dairies. Even into the 1980's, most communities, even small towns, had at least one local, family dairy that provided milk to the community. A large dairy might have had 100 cows that grazed on rolling, grassy pasturelands. Then, in the 80's and 90's, increasing government regulations crippled and shut down countless local, family dairy farms opening the door for increasing industrial consolidation on a massive scale. Like the beef feedlots, dairies became regional operations often with thousands to tens of thousands of dairy cattle housed in relatively small, barren pens. Collectively, these new livestock management models became known as Confined Animal Feeding Operations or CAFOs.

Relative to dairy cattle, beef cattle don't stay in CAFOs very long. Some may have grazed on residual crops in farm fields prior to entering a feedlot, but they spend most of their lives on grasslands that haven't been chemically treated. Lambs are also raised mainly on rangeland and then spend their final days in a feedlot. Swine spend their entire lives on concrete or plastic floors eating grains as do chickens. Most of these animals are terminal, meaning they will not reproduce or pass on their genetics. They live relatively short lives, usually less than 18 months, and serve their purpose of feeding others. There simply isn't much time for pesticides to accumulate in the bodies of terminal livestock.

Pesticides have what is called a half-life, or the amount of pesticide the body excretes through the urine within 24 hours after consuming it. The rest of the pesticide may be excreted later unless the liver and kidneys are too overburdened. In this case, they will be stored in the body. Most American beef, lamb, and swine simply aren't alive long enough to accumulate a significant amount of pesticides within their bodies, in a single generation.

Dairy and egg production are another story. Laying hens, when they are producing optimally, are kept for 2 to 3 years. Dairy cattle are the longest lived of all commercial food animals. A dairy cow's optimal productive lifespan averages around 5 years before they are retired or butchered. Unlike beef cattle, which are usually only fed a concentrated grain diet in the last few months of their lives, dairy cattle and laying hens are fed a concentrated grain diet most, if not all of their lives. In America, dairy cattle, unless grass-fed, are fed a diet consisting mostly of soybeans and corn often supplemented with ground, dried blood or fish meal. Sometimes, this may be supplemented with cottonseed meal or soybean hulls. In 2020, 94% of all soybeans were genetically modified, as was 92% of corn.[2] Almost 93% of American cotton is also genetically modified. It's a diet of convenience and economy. Externally, this diet does not seem to have any ill effects on its immediate consumers, but globally, cattle cancer rates have been on the rise for the past 15-20 years[34], roughly the same amount of time since GMO feed became the norm. Is it possible that we don't have quite as much control over Agrobacterium as we would like to believe?

When agri-chemical companies defended their products from the 1950's until the early 2000's, much of what they had to say wasn't necessarily dishonest in regards to the chemical's safety. Much of the time, these companies honestly and candidly didn't understand the mechanisms and actions behind their chemicals. Besides, the stakes were lower. When chemical pesticides were first introduced, still, relatively pristine environments could absorb poisonous, halogens and other cides or pass them down the line to less saturated environments. There was always a less concentrated environment to spread these chemicals out on.

Ecosystems are wonderfully designed with all sorts of contingency plans and an amazing ability to restore their own health. What a farm soil, animal, or man ingests isn't what makes them sick so much as what they can't eliminate. By the early 1970's, an EPA study showed agricultural chemical contamination not in some American waterways; They found agricultural chemicals in every waterway in America, every river, every stream. Every one.

We know the effect these agricultural chemicals have on insects and plants, but how do they affect livestock as they accumulate, if at all? About 700 different active pesticides are currently in use.[5] These include insecticides, herbicides, and fungicides. Cides are grouped by their method of killing, and there are 2 primary classes or methods by which cides in agriculture kill.

The first class of cides poisons living systems. Usually insecticides, this class impairs or altogether stops vital processes within insects. However, they are not species specific, only dosage specific. Scientists candidly and frequently admit that they don't fully understand how these chemicals work, but we do know they work primarily to disrupt an individual's nervous system, metabolism, digestion, and/or reproductive system. About 85% of insecticides focus on disrupting or shutting down nervous and muscular systems.[6] These are the acetylcholine inhibitors.

Acetycholine is the Paul Revere of the body. All throughout mammalian bodies, there are receptors for acetycholine to relay its essential messages. Acetycholine tells an animal's lungs to breathe, its heart to beat, and its stomach to digest food. They shout to muscles to run or fight when danger is present. Acetycholine receptors form a complex relay system throughout the brain and spinal cord, soft tissue glands, such as the thyroid, voluntary and involuntary muscles, and bone marrow stem cells. When activated, these receptors enable nerve impulse communication and consequently tell muscles when to move. The vast majority of today's insecticides are irreversible, acetycholine inhibitors. Commonly known as organophosphates, this group of insecticides, causes death by overstimulating and exhausting the insect's nervous system.

The remaining members of this first class of cides are honed-in on stopping or altering insect growth, energy metabolism, and digestive membranes to the extent that the insect either dies or is unable to reproduce. In other words, insects that ingest these insecticides stop growing, are lethargic, can't digest their food, and/or become infertile.

The second class of cides isn't intended to poison plants. Their lethality lies in their ability to starve their target plant to death by creating nutrient deficiencies within that plant. This second class includes amino acid inhibitors, fatty acid inhibitors, growth inhibitors, cell membrane disruptors, carotenoid disruptors, and Shikimate Pathway disruptors. The most prominent of the amino acid inhibiting herbicides today is Round-up, but a short list of other modern amino acid inhibiting pesticides might also include the following: Accent, Accord, Amber, Assert, Arsenal, Autumn, Beacon, Cimarron, Classic, Equip, Escort, Express, Everest, First Rate, Glean, Honcho, Matrix, Maverick, Olympus, Option, Oust, Peak, Permit, Plateau, Pursuit, Python, Scepter, Sledgehammer, Resolve, Raptor, Touchdown, and many, many others. While Glyphosate stops tryptophan and phenylalanine production, Chlorsulfuron for example stops the target plant from producing two other essential amino acids, valine and isoleucine. Resistance to these herbicides has reached a level where they are frequently applied as cocktails to enhance their strength. These herbicides not only disrupt amino acid and protein production in their target plants but also in the soil microbiome by destroying algaes and Mycorrhizae. With the algaes and Mycorrhizae diminished or destroyed,[7] fields sprayed with these amino acid disruptors are most likely producing fewer amino acids for livestock to eat today than they would have 80 years ago. Thus, logically, the nutrient deficiency extends beyond the target plant and affects plant consumers. Of course, it's impossible to know for sure whether our livestock today are consuming fewer amino acids than livestock did 80 years ago because we can't turn back time and test soils from 80 years ago and no otherwise comparable soils exist today. Until quite recently, all livestock diets were based around the concept of crude protein. A grossly oversimplified concept, as long as the crude protein or CP of an animal's feed was high enough, we considered that their protein needs were met. We didn't measure any *"other nitrogen...such as (complex) amino acids."* [8] Like synthetic, inorganic nitrogen fertilizers, crude protein was based off the idea that livestock needed a certain percentage of nitrogen in their diets. Now, we recognize that livestock, like plants, don't actually absorb nitrogen very well. Their bodies are searching for amino acids and more complex proteins, and they need them in specific ratios to each other. The closer we can achieve the optimal amino acid balance for an animal, the healthier and more productive that animal can become.

The next group within the nutrient deficiency herbicides are the fatty acid inhibitors such as Harness, Keystone, Sure Start, Surpass, Volley, and Warrant. This typically Chlorine and Fluorine-based group of herbicides stops plants from producing long-chain fatty acids. Fatty acids are basically giant, carbon storage bins in the soil, and they are absolutely essential for neurological, reproductive, and digestive health in mammals. Famous fatty acids include Omega 3's and Omega 6's. Soil nematodes create Omega 3's. These are the same nematodes that Mycorrhizae capture, dissolve, and send the nutrients from back to their partner plant in exchange for complex carbohydrates. Another tiny, soil microbiome creature that creates Omega 3's are Springtails. In good times, Mycorrhizae store fatty acids away both for their own later consumption and also as a means of building soil currency. Healthy soil microbiomes are rich in fatty acids, and we might consider them the foundation to soil health. In fact, some soil scientists are now gauging soil microbiome health by measuring the quantity and diversity of fatty acids within the soil. Like amino acid disrupting herbicides, fatty acid disruptors stop fatty acid production in plants and the soil microbiome.

The next major herbicide group are growth inhibitors. These include 2, 4 D, Dicamba, and Clopyralid. These primarily Chlorine and Fluoride-based herbicides aim to stop a plant's growth at particular junctures. They may stop seedlings from sprouting or roots from growing. Others stop the growth of stems or prevent plant cells from forming cell walls. Some stop cells from dividing and replicating.

Photosynthesis inhibitors as a class are arguably the most poorly understood of herbicide groups. They were believed to essentially starve plants to death by inhibiting photo synthesis. However, the timeline of the plant's response to these chemicals calls this theory into question. A more likely theory is that chemical reactions caused by these herbicides produce free radicals that rapidly damage plants internally beyond repair.[9] Cell membrane disruptor herbicides may actually function similarly to photosynthesis inhibitors in that they promote free radicals in the plant destroying fat and protein membranes.[10] This highly toxic group of herbicides is often used to desiccate fields prior to vegetable production. Valor, Lotus, Inspire, Action, Aim, and Authority are a few of their trademark names. They have become increasingly popular in recent years due to glyphosate resistance.[11]

Pigment inhibitors are often used on rice fields.[12] It is notable though that pigment inhibiting herbicides actually prevent a specific amino acid, tyrosine, from breaking down into Vitamin E and creating carotenoids. The plant needs these carotenoids to protect itself against stress. These herbicides also stop the plant from creating chlorophyll. The plant loses both its pigmentation, and the ability to feed itself or its Mycorrhizae. It dies, and the chemical components disintegrate into the soil or get blown away. A new crop is planted in its place.

Most insecticides, herbicides, and fungicides have federally regulated withdrawal times. In theory, these withdrawal times safeguard livestock and people from eating foods with any cide residues, but we know that livestock are receiving trace amounts anyways. Dairy cattle are likely receiving the most.

When livestock are exposed to pesticides, they may not have the advantages of generational length and prolificacy that insects do, but they are able to disperse any ingested pesticides over exponentially larger bodies. Their bodies first respond to these dangerous chemicals by, when possible, eliminating them through the liver and kidneys and subsequently through manure and urine. A pesticide's half-life is the length of time it takes for half of the amount ingested to be eliminated out of the body through manure and urine. Round-up, for example, has a half-life of about 24 hours. The other half of the toxin may be eliminated later or not at all. Like a busy post office receiving more packages than it has room or personnel to handle, the liver and kidneys can become overwhelmed with more dangerous chemicals than they can eliminate. When these systems become inundated, the body, intuitively, shunts them away to less essential bodily tissues such as the extremities and bone marrow followed by hormonal glands made up of soft tissue such as the thyroid and the pituitary. The goal is to prevent these chemicals from damaging essential organs, such as the nervous system and heart. If the body is continuously assaulted by more incoming cides, then they will continue to accumulate in remote and then increasingly vital parts of the body until the liver and kidneys have a low enough load to process and eliminate them, if they ever do. An animal's age and gender also influence how well and how quickly they are able to detox these chemicals. Male cattle are generally better at eliminating Round-up while females often retain it in their bone marrow and soft tissues. Female bodies will eliminate Round-up over time through their placenta into their offspring and through their milk.

We can measure pesticides in manure and milk from dairy cattle, but we don't very often. A 2019 study comparing conventional to organic milk pesticide residues was the first of its kind. A well-designed study with a diverse sampling of milk from across the U.S., this study found modern day pesticide residues in up to 60% of conventional milk samples. DDT and its byproduct, DDE, were found in almost 100% of milk samples, both organic and conventional.[13] Very few farms in the world have not been affected by these chemicals. DDT is banned within the U.S., but still commonly used in many, if not most countries, globally. Traces of these pesticides show up in dairy cow milk on nearly every continent.

In general, our testing abilities for measuring pesticide residue in milk and eggs are quite poor. Scientists are just now researching convenient methods to test for and measure pesticide levels in these products, and most of this testing and measuring is occurring in countries outside the U.S., such as Europe, the Middle East, Africa, or even China.[14,15] For decades, American agro-chemical companies, along with our research institutions, and the FDA have insisted that bio-accumulation within livestock species was highly unlikely, if at all possible. Thus, how these chemicals may or may not interact with each other outside of their intended insect and plant targets is largely unknown, and while some do not share a common method of action in livestock, most do.[16]

In 1989, the Canadian Veterinary Journal recorded a heartbreaking incident of 16 Holstein dairy cattle that were exposed to Terbufos, a common acetylcholine inhibitor and organophosphate. The Terbufos had been applied to canola seed that was intended to be sown, and the cattle had somehow gotten into it. The signs and symptoms of acute acetycholine inhibitor poisoning can take up to a week to manifest, and they're fairly unremarkable symptoms such as diarrhea, colic, and difficulty urinating. These are combined with central nervous system depression. Have you ever asked a cow if she has a headache? Checked her speech for slurring? Asked her what year it was to see if she was having trouble concentrating? This farmer wasn't aware that his cattle had been poisoned until it was too late for 6 of them. Naturally, the cattle couldn't and rightfully shouldn't have entered the food supply, so it was recommended that they be incinerated.[17] Terbufos is commonly used today both as a soil insecticide and a soil nematode insecticide. It's also commonly now found in dog and cat flea repellant collars.

It cannot be said enough that acetylcholine inhibitors are not species specific; they are dosage specific. What about lower-level acetycholine inhibitor exposure? What might subclinical symptoms look like? A decline in fertility is usually the most notable, long-term subclinical impact. In one study where dairy cattle were exposed to acetylcholine inhibitors, conception rates dropped from 50% to 16%. Progesterone rates also diminished significantly in these females.[18] When a cow fails to conceive, producers don't typically check for potential acetycholine inhibitor exposure. They find a replacement heifer that hopefully will produce a calf, and that cow goes to the auction barn. Ultimately, her meat may find its way into a family's backyard hamburgers at a kid's pool party or a girl's night out at the local taco joint.

We have no reliable, pre-pesticide, pre-synthetic, inorganic fertilizer, pre-GM control soils to use in scientific studies. Even fields that have been fallowed for the mandatory 7 years needed to be certified as organic, can't compare to the fields and soil microbiomes from the mid-19th century or for millenia before that. We do know that insecticides, herbicides, and fungicides have wrecked havoc on our soil microbiome and destroyed both nutrient building and storage mechanisms within the soil. We don't know how much nutrient values have dropped in livestock feed over the past hundred and fifty years or so. We do know that nutrient supplementation improves livestock performance. Whether it be amino acid or fatty acid supplementation, every area of dairy cattle production improves with the addition of these nutrients to an already "balanced" feed ration. Additional amino acid supplementation significantly improves both the volume and the quality of milk a cow produces. It even improves the cow's physical appearance or body condition score.[19] Fatty acid supplementation has been proven to increase female fertility. Cattlemen are transitioning from focusing solely on crude protein to balancing cattle diets for specific amino acids. Animal nutritionists, for decades, used what we now know, were oversimplified feed tags to balance cattle rations. Thus, it is difficult to estimate how different the amounts of complex proteins and fatty acids our soils today are offering our livestock today compared to previous epochs. We also don't know what potential complex proteins may have been built into dairy milk from previous generations with these. While cattle are capable of producing their own amino acids within their digestive tract, they must have adequate building blocks with which to do so. The simple fact is that while we are well-familiar with essential amino acids, beyond this, we know that there exist an unknown number of amino acids with which we are unfamiliar, both of their composition and their long-term effects, not merely for survival, but for fulfilling an individual's full genetic abilities. In short, today we know enough to know there is much we still do not know.

  The one commonality to all CAFO operations is an abundance of manure and the question of what to do with it. CAFOs, in America, produce, an estimated, 133 million tons of manure each year.[20] The smaller, diversified, farms and ranches of the past could easily gather their manure and reapply it to their own pastures, fields, or even a family garden. Rich in organic nitrogen and other essential soil nutrients, manure has historically been viewed as an ideal fertilizer, but most CAFOs today aren't usually close enough to farm fields to make this practical. The sheer expense of shipping manure, which is high in water content, generally prohibits this. To compound this problem, CAFO manure is frequently high in harmful bacteria, animal antibiotics, and pesticide residues. Having been underbid by synthetic, inorganic nitrogen and sidelined by consumed pesticides, CAFO manures sit idly by while their rich nutrients are evaporated into the atmosphere or leach into nearby water systems.

Chapter 7 Dead Zones

"cnid" = Latin: nettle, stinging plant

"A third of the living creatures in the sea died…" ~Revelation 8:9

On a cornflower blue canvas, a yellow, single-engine crop duster with a blue, horizontal stripe running from propeller to tail, ends its pass over young, tidy, green rows and begins its steep ascent and tight turn to make another pass. The pilot carefully scans the land below him to avoid accidentally spraying houses, school buses, or passing cars with pesticide. Across the county, an immense, bright green tractor rolls slowly and steadily along, gently pulling a liquid, fertilizer sprayer behind it. In the past 2 decades, sprayer technology has become incredibly precise. The flow rate, intensity, spray pattern, and angle can all be easily and accurately adjusted, oftentimes from a cab-mounted iPad. Operators can rapidly scroll through farm software-generated spreadsheets, graphs, and maps depicting massive amounts of current and prior field and crop data. At our fingertips lie troves of data the field workers of Leptis Magna couldn't even imagine, yet despite all our impressive, technological advances in precision farming, 15-60% of what we spray won't be absorbed into the plant. How much the plant absorbs depends upon many factors including heat, sunlight, humidity, and the plant's life stage. Some of what we spray or apply will land on the soil and persist there for decades, perhaps even centuries. However, around 60% of the nitrogen fertilizers applied and between 15-40% of pesticides, will either evaporate into the atmosphere or runoff into waterways and ultimately our oceans.[1]

Not far from the skeleton tower whose single, strobing light alerts cargo ships to the Mississippi River's Southwest Pass, they float effortlessly according to the rhythm of the currents. Draped below them are the delicate folds of one of nature's most elegant fabrics. Jellyfish are one of the only living creatures here. Species that rely on oxygen in the water, such as their natural predators, can't survive in this hushed and barren void. This is the western hemisphere's largest dead zone. The size of the dead zone south of the mouth of the Mississippi River varies annually, but it's usually between 2,000-8,000 square miles. Never a grand tourist destination, like the beaches of the Florida Panhandle, it has historically been a sportsman's paradise, supplying seafood for those tourist destinations.

Many of our oceanic dead zone's death certificate's bear the same cause of death as that of our soil microbiomes. Both were fed to death with synthetic, inorganic nitrogen. As the applications of synthetic, inorganic nitrogen fertilizers have grown, so have the number of oceanic dead zones. In 1960, there were 10 documented oceanic, dead zones around the globe. By 2006, there were 169. Today, there are over 400.[2] Dead zones or some level of oceanic eutrophication are found near almost every continent today, including Antarctica.[3] Globally, close to 200 million tons of synthetic, inorganic nitrogen fertilizer are applied each year. If 60% of this is washed downstream or evaporates into the atmosphere, then close to 120 million tons of synthetic, inorganic nitrogen is entering our waterways and skies, every year and has been for decades. Meanwhile, American CAFOs will produce around 133 million tons of manure each year. Most of it will be wasted and also either runoff into waterways or evaporate into the atmosphere.

This bio-accumulative cocktail of partner pesticides and i.v. drip nitrogen was sold to the American people as the remedy to feed a burgeoning, global population, but it has virtually eliminated the fish, shrimp, and oysters that once were a primary source of food and significant source of income for the Gulf Coast. Excess synthetic, inorganic nitrogen not only promotes rapid growth in farm crops, it also causes rapid growth in micro-ocean plant life. Just as with farm soils, this hypergrowth of algaes and plankton causes them to grow to a population density that the ocean carbon to nitrogen ratio cannot support. This exploding population also rapidly consumes all the available oxygen around them and kills oxygen-dependent crustaceans, fish, and turtles. As these animals decompose, further ammonia and death gasses are released into the ocean and then the atmosphere. Marine mammals left swimming in an ocean of death gases either die themselves or flee. Finally, the algae themselves die, leaving only the jellyfish behind gracefully dancing amid the death.

For decades, a false, nitrogen cycle was erroneously believed and taught to multiple generations that plants preferred consuming inorganic nitrogen to organic nitrogen. Like ocean algae or phytoplankton, high levels of inorganic nitrogen cause soil microbe populations to rapidly increase to a point. The microbial population growth stops when it has devoured all of the soil's stored carbon deposits. The imbalance creates, essentially, a dead zone within the soil microbiome, and this is where we've been growing our crops for the past 70 years, a soil dead zone.

Because our plants prefer complex soil proteins and amino acids to an inorganic nitrogen i.v. drip, much of the inorganic nitrogen we apply to our fields isn't absorbed by the plants either. So, it evaporates or washes downstream. Around 120 million metric tons of inorganic nitrogen goes into the atmosphere and the water in and around our planet each year.

Synthetic, inorganic nitrogen is created under high temperatures and intense pressures by combining hydrogen and atmospheric nitrogen to produce ammonia. Initially, the heat and pressure came from coal. Now, today, about 20% of inorganic nitrogen production still relies on coal. About 80% of the production is underwritten by natural gas. Inorganic nitrogen production consumes about 3-5% of all natural gas production globally. Making just one ton of inorganic nitrogen fertilizer requires 33,500 cubic feet of natural gas.[4] The process to create synthetic, inorganic fertilizer also creates more carbon dioxide (CO_2) than any other process in the world. In 2010, synthetic, inorganic fertilizer production created around 450 million tons of carbon dioxide.[5] Much of the production is actually beneficial to the oil and gas industry, because it often puts "stranded gas" or gas located in parts of the world without economical markets, to use. Turning the gas into easily transportable, though still highly volatile, synthetic, inorganic nitrogen fertilizers has long been considered a win-win for both the oil and gas and agricultural industries.

Creating synthetic, inorganic nitrogen also requires substantial amounts of water. Ironically, excess water is perhaps the primary reason why its primary competitor, CAFO manure, is rarely used as fertilizer, even though it is organic and preferred by the soil. CAFO manure's high-water content makes it too expensive to ship. Not to mention, most CAFO manure is tainted with an amalgam of harmful bacteria, pesticides, and antibiotics. Thus, millions of tons of both inorganic and CAFO nitrogen find their way into our waterways and oceans every year.

Seventy-eight percent of our atmosphere is naturally comprised of a form of inorganic nitrogen. It serves to naturally protect the earth's atmosphere, but the excess nitrogen that evaporates from our farm fields is in a different form. It doesn't protect us, and it can't simply return to where it came from. What happens to it is largely unknown. While carbon dioxide is the focus of most atmospheric discussion, it is *"the nitrogen cycle has been altered more than any other basic element cycle."*[6]

For nearly a century, we have drenched our fields and soils in synthetic, inorganic nitrogen and pesticides. After having been applied, these chemicals once diffused across the environment. Today, these diffusions are bioaccumulating and becoming increasingly concentrated on the global stage. When the problem washed down the creek and into the river, it was diluted to the point that we felt safe to swim, kayak, and fish in it. When it was washed down the river and into the ocean, it was diluted once more, and we felt fine to fish, swim, soak, and let our babies play in it. The problem was out of sight and out of mind. Because we didn't have a solution, we were going to ignore the problem. That's becoming and will continue to become an increasingly difficult thing to do.

Big skies and vast, grassy plains join up with distant, verdant, pine-covered mountains. A shallow, rocky river lazily winds its way across the vista. Scrappy, little horses restlessly graze in herds across the valley. Smoke rises from a white, circular yurt while joyous children play outside. The Mongolian people aren't tillers of the soil. They have maintained their nomadic lifestyle for millennia. Doubtless, they never dreamed that their peaceful valleys would be softly and quietly laden with pesticides, but they are.

The LRAT is short for long range atmospheric transport. Up to 20% of all synthetic, inorganic nitrogen and 40% of pesticides applied to crops don't wash down waterways. They evaporate into the atmosphere. Although the spread of pesticides and synthetic, inorganic nitrogen across the atmosphere isn't talked about much, Chinese scientists are now studying how pesticides are picked up into the atmosphere and then deposited elsewhere, oftentimes, far, far away from the farm fields. [7]

"Wearing a mask can't block the feeling of eating dirt." ~Chinese, Weibo User[8]

For China, the problem may be more immediately apparent than in America. Sandstorms have been plaguing the country for years, and they seem to be growing in intensity and frequency. Nicknamed the *"China Dust Express,"* these sandstorms don't only affect China. The neighborhood watch of Japan, North Korea, and South Korea have all voiced their complaints. While the eastern U.S. and the Caribbean occasionally receives a dusting of Saharan sands or enjoy a red hazy sunset, California, western Canada, and western Mexico likewise enjoy free, golden skies, compliments of China. The skies aren't just filled with migrant sand though, intermingled with them are synthetic, inorganic nitrogen and several families of cides.

Once pesticides have been lofted up into the atmosphere, there are two ways for them to return to earth. One is through dry deposition where chemicals simply fall back to earth. The other is through wet deposition. This is when pesticides fall back to the earth in rain, fog, or snow.

Whether pesticides or inorganic nitrogen, these chemicals are seldom carried away on the winds for a one-time trip to make a new, forever home somewhere else. They are typically jetsetters. They might travel twenty miles down the road from the farm field where they were first supposed to go to work to alight on a schoolyard playground. From the playground, they may be lofted up again and carried higher and further across the continent. This cycle may continue indefinitely until they are either absorbed by a plant or consumed or inhaled by an animal or human. Atmospheric currents don't respect international boundaries. There go where they like, and they like to travel. They've been translocating pesticides and fertilizers globally since we began applying them. As early as 1969, 3 out of 5 snow samples from Antarctica tested positive for DDT.[9] Some of the halogen-based pesticides found traveling the world, have been doing so for close to 100 years now. Ice samples from Antarctica to Svalbard, Norway, contain halogen-based pesticides. Water, soil, and air samples from the Tibetan plateau and the Himalayas do as well.[10] When happy, carefree hikers deeply inhale fresh, mountain air or refill their water bottles from pristine mountain streams, they are likely inhaling and drinking in agricultural pesticides. Pristine environments don't exist anymore, and neither do pristine farmlands.

More recently developed pesticides aren't any better at staying down home on the farm. Glyphosate may break down into aminomethylphosphonic acid or AMPA, a slightly less toxic version, or it may not. Researchers have found it persisting in the soil up to 10 months after being applied.[11] Ten months wasn't the amount of time it took for glyphosate to disintegrate; it's just when the researchers ended their study. While AMPA tends to be more of a homebody, glyphosate is happy to see the world.

Few studies measuring Glyphosate in the atmosphere are available. It takes courage to ask these uncomfortable questions, but living in ignorance isn't helping anyone. For those brave enough to do the research, the results consistently confirm that Glyphosate bioaccumulation is increasing in the air we breathe. A 2014 study, in Northeastern Brazil, found 100% of air samples, in both rural and urban areas, contained Glyphosate.[12] A 2018 study, in sparsely populated Argentina, found halogen-based pesticides and Glyphosate in more than 80% of the rain samples taken. Their local soils were determined not to be the sources of these pesticides.[13] A 2011 study, found between 60-100% of air and rain samples from Iowa, Mississippi, and Indiana contained Glyphosate.[14]

"It is out there in significant levels. It is out there consistently."
~Paul Capel, USGS[15]

The Hauts de France region that stretches from Paris to Calais is a surreal blend of tourist perfection for the upwardly mobile. Charming villages dotted with chateaus and fairy-tale castles are interspersed with miles upon miles of family hiking and biking trails through eden-like forests, pasturelands, and orchards. Michelin-starred, locally sourced, artisan, and farm-to-table restaurants abound. It's also the home of a UNESCO biosphere reserve. On a clear day, from the sheer, grassy cliffs of its coastlines, one can see the White Cliffs of Dover. You might never suspect that in 2004, 14% of 59 air samples taken in this region detected Glyphosate in the air, at breathing level.[16] That was almost 20 years ago, and Glyphosate usage hasn't slowed down since. Like the tourists that frequent the area, the Glyphosate in the study didn't appear to originate at nearby farms. It seemed to have drifted in from elsewhere.

Glyphosate generally becomes airborne when mixed with dust particles and returns to the earth through heavy rain falls. Unlike a conscientious, crop-dusting pilot, rain falls wherever it pleases, including over homes, playgrounds, and hospitals. One of the greater scientific unknowns is how all of these transgenes, nitrogen, and pesticides interact with each other as an environmental cocktail. Little to no research has been done on such interactions and their long-term ramifications.

Global dimming is the measured decrease in the amount of sunlight reaching the earth. The strength of the sun, likely, hasn't changed over the years, but the amount of sunlight that is able to reach and warm our soils has varied substantially. Global dimming occurs when large amounts of tiny airborne particulate gather in clouds thus blocking the sun and reflecting sunlight back into space. Are some of the inorganic nitrogen and pesticides, now filling our skies, contributing to global dimming? Quite possibly. It might sound like a cure for global warming, but it isn't healthy for our soil microbiome. It also slows down new topsoil creation. The bacterial to fungal ratio of healthy soil microbiomes is a reflection, among other factors, of the amount of sunlight and heat the soil receives. Decreased sunlight and decreased heat slows down soil microbial metabolism, energy consumption, and reproduction. These are what drive the creation of new topsoil. Consequently, less sunlight means slower soil production. When you are already losing soil at an unreplenishable rate, slowing down new soil creation doesn't seem helpful.

Chapter 8 Apex Omnivore

"I was sitting in my house a couple of weeks ago when my doorbell rang...it's a different feeling when your doorbell rings today opposed to 20 years ago. Twenty years ago, your doorbell rang? That was a happy moment in your house. It's called company....Now, your doorbell rings?...Get...down! Somebody get the sword in the living room! Somebody is outside! I think they saw movement!"
~Comedian Sebastian Maniscalco

Weston A. Price was a dentist who studied the diets of indigenous people and their dental health in the early 20th century. The indigenous peoples he studied remained largely isolated from outside influences, and most had nearly perfect dental health, even into their old age. Their teeth were naturally straight and white with little tooth loss. Based on his studies, he contributed their excellent health to their primarily animal-based diets. These diets were rich in fat-soluble vitamins, A, D, E, and K and complex proteins. Today, the Weston A. Price Foundation carries on his work, and they espouse that what a person eats or doesn't eat carries down to the 3rd generation. They believe that the food you eat today will influence your own epigenetic code and even the epigenetics of your grandchildren. For Baby Boomers who were raised, sometimes literally, playing in DDT, the belief that their consumption patterns didn't affect them personally, was likely true. They had inherited the nutrition of their grandparents. Their grandparents were known as the Lost Generation. They were born between 1890 and 1915, when the only pesticide known was arsenic and when horsepower generally still referred to horses. Beef and pork might have been locally grown or shipped-in on refrigerated railcar. Vegetables and fruits were predominately locally grown, even if staples and spices were imported from other states. The Lost Generation's children, the Greatest Generation, inherited this goodness and passed it along to their children, the Baby Boomers. It was during this era, the era that Baby Boomers were born into that the Green Revolution, the widespread use of synthetic, inorganic nitrogen fertilizers, and dispersal of halogen-based pesticides began.

If Weston A. Price's theory holds true, then even the children of Baby Boomers might have retained genetic and nutritional buffers that they inherited from their Greatest Generation grandparents that would shield them against disease and dysfunction. If Weston A. Price's theory holds true, the grandchildren of Baby Boomers would be the first to evidence signs of mass toxicity. This generation is Generation Z.

As the body attempts to detoxify itself from increasing levels of toxicity, it will fight for survival. As with other mammals, the human body first eliminates as many poisons as possible out of the body through the digestive system, liver, and kidneys. The ability of the digestive system, along with the liver, kidneys, gallbladder, and pancreas to detoxify foreign chemicals from the body depends upon having more than adequate nutrition. A loaded detoxification system needs more resources, such as vitamins, minerals, and antioxidants, to complete its job, not less. Unfortunately, due to the condition of our soil microbiomes, most people's detoxification systems don't have enough nutrients to adequately process these inherited chemical burdens. As with other mammals, when the body's detoxification systems are loaded to their maximum capacities, the body will naturally prioritize its own survival and protect the central nervous and cardiovascular systems by shunting poisons to less essential areas of the body, such as the extremities and bone marrow. Like empty Amazon boxes piling up in a closet, without an outlet, these toxins will eventually accumulate and spill over into essential bodily systems. Females have an additional mechanism for detoxification. The overburdened female body can disperse excess toxin concentrations through childbirth and nursing.

"Gluten-Free," "Leaky Gut," "Brain Fog," "Chronic Fatigue," "AIP Protocol"—these terms were virtually unheard of prior to the mid-2000's. By the mid 2010's, they were everywhere. Females in the Generation Y and Millennial range seemed to be the predominant sufferers of these very, unscientifically labeled health crises. In fact, many physicians didn't believe these conditions were real. Perhaps, it is because females are generally, smaller-bodied than men and were more sensitive to smaller concentrations of chemicals. Perhaps, the female hormonal system, being more complex, is more quickly disrupted than the male system. Many women were told that their symptoms were all in their heads. Initially viewed largely as pseudo-science, functional medicine physicians began treating for general "inflammation" and focusing primarily on the digestive system. The reality was that it was not pseudoscience. The inflammation response in consumers was, perhaps, mirroring the inflammation response that scientists had measured in plants in response to pesticides.[1] Additionally, physicians were now treating transgenes in digestive systems which had never existed before. Standard, diagnostic tools couldn't find what they weren't built to find. Diagnosticians couldn't identify what they weren't trained to identify. Of course, pesticide usage and agricultural transgenes could be ruled out as contributing factors. Science had proven they were perfectly safe...or had it?

The reality was that the research performed on the genetically modified or trans crops and chemicals people were consuming at unprecedented amounts had been mostly scuttled.

> *"But...internal records were made public due to a lawsuit and the deception came to light. The agency's (FDA) newly released 44,000 pages revealed that government scientists references to the unintended negative effects.... were progressively deleted from drafts of the policy statement (over the protests of agency scientists. They further revealed that the FDA was under orders from the White House to promote GM crops and that Michael Taylor, Monsanto's former attorney and later its vice president, was brought into the FDA to oversee policy development. With Taylor in charge, the scientists' warnings were ignored and denied."*[2]
> ~ *"Genetic Roulette"*, Jeffery M. Smith

Consumption of genetically modified or trans foods remodels the human digestive microbiome at the genetic level, producing new and harmful digestive microbes. Some of the microbes may have the ability to reproduce pesticides within the human digestive tract, effectively killing off the native microbes. After doing so, they may establish transgenic control and dominance over the bacterial flora of the digestive system.[3] Glyphosate also weakens the cellular junctions of the digestive tract and allows partially digested foods to slip through into the bloodstream. Once inside the bloodstream, these foods are viewed as foreign bodies and rejected thence forth from the entire body, and a new allergen is born. These weakened intestinal cells further create a heavy metal highway into the bloodstream which transports them to the soft tissues[4] such as the brain, thyroid, and reproductive glands. Glyphosate also weakens mitochondrial membranes, disrupts ATP synthesis, and causes oxidative stress to the mitochondria, the energy production plant of the body.[5] In other words, glyphosate drains energy from the body at the cellular level.

Not only were women's diets being saturated with trans-genes, herbicides, and insecticides, their foods were grown in increasingly nutrient-scarce soil microbiomes. Once Mycorrhizae were destroyed, plants were left without anyone to hunt down and bring home omega 3's and omega 6's within the soil. Nematodes lost a key predator, and pesticides took the place of native Mycorrhizae to control nematode populations. The result was nematode obliteration and fewer essential fatty acids in the soil and consequently, the plants grown in that soil. The female reproductive system practically runs on fatty acids. Suddenly, estrogen dominance, low progesterone, and PCOS, formerly an almost unheard of diagnosis, became common place. Depression, in general, rose rapidly in Americans between 2005 and 2015, but it rose most rapidly amongst young women. A multitude of cultural and societal changes in our world during that timeframe might have contributed to these conditions, but depletion of dietary omega 3's and 6's are undoubtedly key contributors.

"Gluten-Free," "Leaky Gut," "Brain Fog," "Chronic Fatigue." Disrupted nervous systems, porous digestive membranes, decreased energy metabolism, and dysfunctional reproductive systems---suddenly, the crazy ladies didn't sound so crazy anymore.

Generation Z and Gen Alpha are the first generations to be introduced, in utero, to the effects of this perfect storm of a womb surrounded by an overloaded detoxification system and fed by an already nutrient deficient body while absorbing transgenes, pesticides, and heavy metals. Few intergenerational studies have been conducted on genetically modified or transfoods, and yet all of these factors influence epigenetic changes that take place in the uterus. Some of these changes alter the DNA for generations to come; some may permanently shift a genetic line towards dysfunction creating an incalculable number of domino effects across generations and all of them unhealthy.

Human beings are most vulnerable to harm from pesticide exposure in utero, shortly after birth, and at puberty. In 2006, an EPA report from the Office of the Inspector General reported that *"certain pesticides easily enter the brain of young children and fetuses and can destroy cells. The report concluded that the EPA lacks standard evaluation protocols for measuring the toxicity of pesticides on developing nervous systems,"* and *"also charged that risk assessments cannot state with confidence the degree to which any exposure of a fetus, infant, or child to a pesticide will or will not adversely affect their neurological development."*[6]

Autism, ADHD, depression, severe anxiety, gender dysphoria, social isolation and hostility are a few of the early hallmarks of Gen Z and Gen Alpha. Notably, these are all either neurological dysfunctions or hormonal imbalances, many of which can trace their origins to impaired digestive and detoxification systems and still further back to dying and poisoned farm soil microbiomes.

Incredibly, only a few years past, researchers discovered an entirely new bodily system, known as the endocannabinoid system. This system helps to calm the nervous system.

"The endocannabinoid system (ECS) in the brain primarily influences neuronal synaptic communication, and affects biological functions—including eating, anxiety, learning and memory, reproduction, metabolism, growth and development—via an array of actions throughout the nervous system."[7]

The endocannabinoid system modulates neurogenesis, is neuroprotective, aids in synaptic plasticity, and improves memory processing. It decreases inflammation and stress. It plays a key role in how the brain handles pathological pain and aging. The endocannabinoid system relies upon fatty acids such as arachidonic acid and anandamide. Arachidonic acid, together with another soil fatty acid, DHA, actually make up about 20% of your brain's dry weight. Unsurprisingly, children with autism consistently have low arachidonic and anandamide blood plasma levels.[8] When children with autism have been supplemented with high levels of arachidonic acid, their social interaction abilities have improved substantially.[9]

Arachidonic acid is found in meat, poultry, and seafood, but apparently, these sources are insufficient because companies are now genetically engineering, a natural soil fungus, called Mortierella to produce arachidonic acid in a lab.[10] Similar to Mycorrhizae, Mortierella is a root loving fungus and essential member of the soil microbiome that naturally produces high levels of arachidonic acid in the soil.

Pythium ultimum is a much hated plant-pathogenic, root eating fungus. It once was feared for eating crop plant roots and destroying fields of corn and potatoes, but no worries. Today, we have a whole host of pesticides that will kill it, such as Pageant, Segway, and Adorn, to name a few. Like Mortierella though, scientists, as of late, are trying to grow it intentionally, in a lab. Why? So they can sell the naturally high quantities of arachidonic acid it produces.

In reality, there are a whole host of interconnected pathways between our besieged soil microbiome and our dinner plates that have been broken down and that have led to skyrocketing Autism rates. Glyphosate is designed to stop the Shikimate Pathway that produces tryptophan and phenylalanine. Would it be surprising then to learn that a 2013 study of 138 children with autism when compared with 138 children without autism had relatively decreased levels of the essential amino acids tryptophan and phenylalanine?[11]

The same fatty acids that are low in children with Autism are also low in children with ADHD.[12] As with Autism, there are multiple soil microbiome related pathways that can contribute to ADHD. Acetycholine inhibitors are members of the nerve gas family. As such, they overstimulate nerves, causing them to fire repeatedly to exhaustion. When administered to rats, acetylcholine inhibiting herbicides, true to their nature, created ADHD-like behaviors. Interestingly, researchers found that they could use a receptor in the endocannabinoid system to turn down ADHD behaviors.[13] Acetycholine inhibiting herbicide exposure is currently the primary suspect for ADHD development.[14]

From 2007 to 2019, suicide rates for youth ages 10-23 rose 60%, according to the CDC. Gen Z is reportedly twice as likely to battle feelings of hopelessness and despair as Millennials and three times as likely to say *"that their challenges are so severe that they would be better off dead."*[15] To give a generational perspective, the American Psychological Association reported when asked to rate their psychological health, 74% of both the Greatest and Silent Generations ranked their mental health as very good or excellent. For Baby Boomers, 70% likewise ranked their mental health as very good or excellent. That number dropped to 51% for Gen X and 56% for Millennials. Only 45% of Gen Zer's said their mental health was very good or excellent.[16]

The common explanation for these symptoms has been that Gen Z is facing greater cultural, societal, economic, and technological changes than previous generations, and the challenges they face are unquestionably unique in world history. However, the Greatest Generation and the Silent Generation survived considerable challenges of their own. They survived the Great Depression, the Dustbowl, and World War II, and yet they ranked their mental health as almost twice as good as Gen Z. Something seems amiss. Could it be said that previous generations were able to meet their own challenges with greater resilience because they emerged from childhood without having been starved of essential fatty acids, complex proteins, amino acids, vitamins, and minerals or saturated by pesticides and trans-genes designed to disable their nervous systems?

If our Glyphosate tolerant crops have been engineered to survive without tryptophan, tyrosine, and phenylalanine and if humans were deficient in these amino acids, what symptoms would we see?

> *"She began to cry inconsolably and described her emotions as being 'out of control.' She said that she did not know why she was crying but could not stop. She also described psychic anxiety, difficulty concentrating, loss of energy, loss of self-confidence, and a sense that nothing was worthwhile. She felt as if all the gains she had made over the past few weeks had 'evaporated.'"*
> ~ Simon Young, PhD, referring to a Tryptophan deficient participant in the Acute Tryptophan Depletion Study [17]

Both serotonin and melatonin production rely on the essential amino acid tryptophan as a precursor. The interactions between serotonin and the body are complex, but low serotonin levels are associated with suicide and depression. Not everyone who is deficient in tryptophan experiences depression, and researchers are not certain why, but it certainly is a causative factor in some people's crippling depression and suicidal feelings.

Glyphosate not only inhibits the Shikimate Pathway in plants, but also in the beneficial bacteria found in human digestive tracts. This gives harmful digestive bacteria a competitive advantage, and this destruction of healthy, digestive microbes and supplantation with harmful ones is believed to be another key foundational pathway in depression and anxiety. [18]

Acetylcholine inhibiting insecticides may also play a significant role in depression and anxiety. Imaging studies have shown elevated acetycholine levels in actively depressed patients.[19] It's important to remember that acetylcholine inhibiting insecticides stop the natural give and take of acetylcholine between the nerves, leaving them overstimulated.

> *"These results are remarkably consistent across species, and suggest that blocking or lowering acetylcholinesterase activity, increasing central acetylcholine levels, or stimulating specific cholinergic receptors, all lead to depression."*
> ~ Stephanie Dulawa, Ph.D. and David S. Janowsky, M.D. [20]

Another cause of Gen Z and Gen Alpha's depression and anxiety may be the disruption of the dopamine center of the brain, the substantia nigra. Rats, exposed to Glyphosate during pregnancy, gave birth to young that demonstrated a wide-variety of cognitive abnormalities, including abnormalities in the dopamine system. [21] Another similar study found oxidative stress and depressive-like behavior in the Glyphosate exposed offspring.[22] Yet another study found that young, male rats who were exposed to Glyphosate, in utero, had significant decreases in peptide production in the substantia nigra. Their brains were creating about 50% less of the protein than normal. Known as dynorphins, these neuroproteins are involved in pain, mood, and addiction regulation. They also inhibit excitatory neurotransmissions. Gen Z and Gen Alpha are truly feeling and experiencing life differently than previous generations. These studies all focused on subclinical or subchronic levels of Glyphosate exposure to the mother through drinking water, not overt contact.

Low levels of dopamine are also linked to ADHD and obesity. Children with low dopamine simply eat more to release more dopamine in their brains and feel better because their brains have been underproducing dopamine since before they were born. It's ironic that pesticides and trans-genes were marketed as being necessary to feed a burgeoning world population, and yet the youth they feed are starved of the essential amino acids, they need to thrive. These chemicals have created, in consumers, an insatiable desire to eat more calories than are needed, simply to balance their brain chemicals.

Childhood obesity has tripled since 1993.[23] Like ADHD, anxiety, and depression, obesity has many layers. Both in utero and in childhood, critical nutrients must be consumed at critical times or else the body will suffer lifelong effects, some of which may even pass on to future generations through epigenetics. For example, we know that when critical nutrients are withheld in cattle, at specific growth points, the adult animal will naturally have a higher fat to muscle ratio than in animals who did not have their diets restricted at these critical growth points. A considerable number of studies have correlated subclinical pesticide consumption with shifts in glucose metabolism. While no direct link between pesticides and childhood obesity has yet been found, isn't it possible that simply by consuming diets deficient in a variety of key nutrients, at key growth points, we are causing the childhood obesity epidemic? Are children overweight today because they are actually starving? Do Gen Z and Gen Alpha children simply have to eat more to meet their basic nutritional needs because their foods are so depleted of nutrients?

"The biggest criticism leveled at colleagues born between 1995 and 2007 is that they are self-centered and unable to negotiate or compromise, which inevitably leads to conflict. Other generations often find it hard to understand them, let alone work with them....

> *Some leading thinkers suggest that this generation needs to be controlled and managed, screened out, even, from organizations that depend on teamwork."*
> ~ *Remy Blumenfeld, Forbes Contributor*[24]

Aggression, hostility, and anti-social tendencies are perhaps some of Gen Z's most notorious generational traits. Interestingly, these same traits also seem to have taken early GM or trans-gene food researchers by surprise as well.

> *"Researchers also noted, 'a high level of anxiety and aggression' in both the 'females and young pups from GM groups... They attacked and bit each other and the worker.'"*[25]
> ~*"Genetic Roulette", Jeffery M. Smith*

Why these rat pups fed a genetically modified diet became hostile to each other remains unclear, but we do know that Glyphosate's and herbicidal trans-genes destruction of beneficial, digestive microbes opens the door for harmful bacteria such as Clostridium and Salmonella to thrive. These harmful digestive bacteria produce neurotoxins that can pass through the digestive lining and travel to the brain. Once in the brain, they may stimulate high levels of dopamine which causes oxidative stress and may lead to addiction, aggression, and poor impulse control.

Acetycholine inhibiting herbicides likely play an even larger role in Gen Z's aggression than trans-genes. Researchers now believe that acetycholine receptors may be the mother board for all upstream behaviors that drive this aggression such as mood, inflammation, sociability, impulsivity, anxiety and fear. There are three primary types of aggression, offensive, defensive, and predatory aggression, and acetycholine receptors can influence control over all of them.

Among the Silent Generation, less than 0.05% identified as transgender. For Baby Boomers and Generation X, around 0.2-3% of the population identified as transgender. That percentage more than tripled in the Millennial generation with 1.0% identifying as transgender. Today, 1.9% of Generation Z identifies as transgender.[26] The explosive growth in Gen Z's identification with transgenderism walks hand-in-hand with feelings of gender dysphoria. Up to half of those diagnosed with gender dysphoria have co-occuring diagnosis of depression and anxiety. Between 32-50% of those diagnosed with gender dysphoria have attempted suicide. [27] It is widely-assumed that these co-existing feelings of depression and anxiety, along with suicidal attempts, are caused by a lack of societal acceptance, but what if something altogether different is going on?

"I've done a lot of studies throughout my career which extends back to 1973. This is the very first time that what I've found scared me."
~ John Woodling, University of Colorado researcher [28]

Boulder Creek runs from just south of Winter Park, Colorado in the Rocky Mountains through the town of Boulder and meanders down to the flat farmlands of eastern Colorado. In 2004, scientists made a startling discovery that fish within the creek were displaying characteristics of both male and female genders.

This was an entirely new phenomenon for researchers from the University of Colorado at Boulder, but it is also an easy one to explain. An abundance of estrogens and estrogen mimickers in the water were causing male fish to display female sex characteristics. These estrogens and estrogen mimickers were found by research teams in 80% of the streams they tested in 30 states.

Herbicides, insecticides, and fungicides have powerful influences upon human hormones as well. You could probably fill a small, home library with books about each one. A short list might go something like this. 2, 4 D causes *"synergistic androgenic effects when combined with testosterone."* Acephate disrupts *"hormone expression in the hypothalamus."* Acetochlor *"interact(s) with uterine estrogen receptors and alters thyroid hormone dependent gene expression."* Alachlor *"bind(s) to estrogen and progesterone receptors."* It also interferes *"with the production of enzymes responsible for steroid hormone metabolism."* Aldicarb inhibits *"estradiol and progesterone activity."* Aldrin *"binds to androgen receptors."* Atrazine *"inhibits androgen production, has a weak estrogenic effect, disrupts luteinizing hormone and prolactin levels, and increases estrogen production."* It also *"damages the adrenal glands and reduces steroid hormone metabolism."*[29] That's just the "a's" from a short list of 105 endocrine disrupting pesticides. This list goes on and on and on. A few of these pesticides are now banned in the U.S., but most are still actively being applied to farm soils. Most will remain persistent in our soils, waterways, and atmosphere, for years, if not decades to come.

Gen Z and Gen Alpha's gender confusion is real. Their bodies biochemistries are completely haywire. They have been exposed to untold amounts and combinations of these cides and trans-genes since they were conceived. FDA approval of these cides was carried out on the individual level. There are no generational studies with test subjects receiving daily subclinical doses of these cide cocktails and trans-genes over the past 100 years. We are the test subjects. We are witnessing the effects in real time before our very eyes. Gen Z and Gen Alpha's nervous systems have been respectively stimulated to the point of near exhaustion by chemicals that were designed to do exactly that. Their endocrine and reproductive organs have been receiving false gender messages since before they were born. It is no wonder that Gen Z and Gen Alpha are starting to believe what the cides are telling them. The worst for Gen Z and Gen Alpha, though, may be yet to come.

Around the time GM crops first appeared on the market in 1999, a study emerged that showed rats fed genetically modified feed were sterile by the 3rd generation, if not the 2nd. Another study at Baylor College of Medicine (found that) *'Rats in our animal facility neither breed nor exhibit reproductive behavior when housed on (GM) corncob bedding.' Tests on the material revealed...compounds...that disrupted endocrine function in male and female rats."*[30] Nearly 25 years down the road, we can say definitively that more than 3 generations of other mammals, who were raised on GM feed, have survived and reproduced, but we must bear in mind that like insects, most animal species have shorter generation times and more offspring than humans. This increases their survival ability as a species because surviving offspring with even slightly superior detoxification genetics will intensify those genes when they reproduce with other survivors. Since we've yet to reach our third generation that has been inundated with trans-genes and cide cocktails starting in the womb, we don't know what the cumulative effects will be for the human race. The first generation to consume genetically modified foods, from utero, was Generation Z. If Weston A. Price's theory is correct, Generation Z's grandchildren would be that 3rd generation.

If males have an innately superior ability to detoxify their bodies to females, does this mean that they are unaffected by less nutrient dense foods, trans-genes, and cide cocktails? The short answer is no. Between 1973 and 2018, men's sperm counts dropped by 1% a year. Between these years the average man's sperm count dropped by 52% globally.[31] Worse yet, declining sperm counts are accelerating. That 1% a year average decline increased to 2.64% in 2020. Those aren't sperm count declines from men struggling with infertility. They represent the normal, average man. The numbers are troubling not simply from a reproductive standpoint, but because men with lower sperm counts generally have a shorter life expectancy. Testosterone rates among men are also declining, in general. Testosterone level decline is typically associated with male aging. These results are consistent across male age groups. When declining sperm counts and declining testosterone levels are paired together, we may be looking at a generation of men who will age more rapidly and live shorter lives than they were once genetically capable of having.

Round-up seems to affect men differently than women not only during and after puberty, but also in utero. Round-up exposure during rat pregnancy produced merely a delay in reproductive development in female offspring, but in male offspring, the effects carried into adulthood. Male offspring, who were exposed to Round-up in the womb, had lower testosterone levels in general than unexposed male offspring. They also produced fewer daily sperm at puberty, and the sperm they did produce were often abnormal.

In another study, when mature rat testis were exposed to very low-level Glyphosate concentrations in a lab, testosterone dropped by 35%. [32] Perhaps the manliest thing a man can do in 2023, is reject Round-up.

Between 2014 and 2017, the percentage of Americans, between the ages of 55 and 64, who were diagnosed with early on-set Alzheimer's or dementia grew by 143%. During the same time frame, early on-set Alzheimer's increased by 311% for 45-54 year olds. For Americans between the ages of 30-44, early on-set Alzheimer's or dementia diagnosis grew by 373% in a 3 year time span.[33] Why is early on-set Alzheimer's increasing at such an alarming rate? Unsurprisingly, there is an entire web of connections between Glyphosate based cides and Alzheimer's. These include the hijacking of digestive microbes and direct attacks on the substantia nigra, the dopamine center of the brain.

The Mississippi isn't just a highway for topsoil loss; it's the borderlands for some of the highest rates of national depression,

There Might Be Hope

Alzheimer's, and Parkinson's disease in America. In 2021, Mississippi and Alabama ranked 1st and 2nd in the nation for Alzheimer's death rate. Arkansas and Louisiana rank 5th and 6th respectively. Sandwiched between them are Washington and Georgia, also states known for their agriculture.[34] Multiple studies are now showing that there is a relationship between the air we breathe in and neurodegenerative disease. Specifically, there is a relationship between breathing in fine particulate matter and inorganic, nitrogen fertilizer in the atmosphere and developing Alzheimer's, dementia, and/or Parkinson's.[35] [36] Could some of this fine, particulate matter be a cide or an amalgam of cides? Quite possibly.

When the cascade of generational effects is viewed altogether, we have to question whether our aim is really to exterminate insects and weeds or our entire civilization. If we continue on our present course, we won't have to worry about whether or not we have the ability to feed a planet of 8 billion people anymore. We won't have 8 billion people to feed. If current trends continue, a substantial percentage of our population won't have children and will be sidelined from life, at a young age, with early on-set Alzheimers. If a plan had been hatched to conquer America without firing a shot and without ever being held accountable, while making a fortune, it could not be working any better.

Chapter 9 - The Forests of Arabia

"In the forest in Arabia you will lodge..." ~Isaiah 21:13

Perhaps no country on earth is as zealous for agricultural and soil restoration than the reigning king of Middle Eastern oil, Saudi Arabia. Saudi Arabia is predicted to be the last oil producing nation in the world, but current leadership is wisely choosing to focus their attentions on the cultivation of something of a modern-day miracle. They want to transform their vast deserts into lush forests and fertile farmland.

It's perhaps difficult to imagine the vast peninsula as anything other than desert, but like the Sahara and Leptis Magna, barren, sand dunes and camels, campfires and coffee don't represent the Saudi Arabia of old. To being with, the first coffee trees didn't arrive on the Arabian Peninsula until about 15th century A.D. Moreover, although it has likely long been an arid country, it seems doubtful that Saudi Arabia was always desert. Fossils of hippos and elephants, *"much larger than the African elephant"* have been found there.[1] Historical evidence shows that while the inhabitants of this area grew rich through cross-continental trade of metals and spices, they could hold and did hold their own agriculturally. The earliest known inhabitants of Saudi Arabia were likely not a mere congregation of nomadic tribes, but developed, landed kingdoms and trading centers. Some even point to the possibility that Job, *"the greatest man of the East,"* with his 500 donkeys and 500 head of oxen for plowing, may have been from Northern Arabia. The Bible lists 3 of his friends as hailing from city names that were known cities on the Arabian Peninsula. Prior to being conquered by the Assyrian Empire, evidence seems to support a semi-arid habitat with grassland valleys where shepherds grazed their flocks. Becoming nomadic was likely a survival mechanism the Arabian people wisely learned to graciously embrace and now can recall with pride rather than an initial preference. Like the Sahara, the Arabian desert appears to be more desertified than natural desert. Recent excavations have even found thousands of what closely resemble, massive livestock gathering and herding chutes, called 'kites.'[2]

Saudi Arabia is attempting to create a different kind of 'kite' today. Ambitious, motivated, and well-funded, the Saudis are exploring the best means to reclaim the fertility of their soils and possibly even, cool their climate. Following China's lead, they want to plant walls of trees that will capture water, rebuild the soil, stop the sandstorms, potentially restructure weather patterns, and reclaim the desert. It's called the 1 Trillion project, and the goal is to plant 1 trillion trees before 2030. The hope is to reclaim desertified land with forest. It sounds idyllic. It sounds Eden-like. It's not working. As anyone who has ever planted a tree knows, simply planting a tree doesn't guarantee its growth. Trees need care. Trees need water. Trees need a healthy, soil microbiome.

When it comes to healing their soils though, Saudi Arabia likely has some considerable advantages in its pursuit that most other countries don't have. Namely, they aren't battling soils saturated with cides and trans-genes to the degree that other countries are. As the top livestock producer in the Middle East, they also have dairy cattle, sheep, goats, and horses. Thus, they have the advantage of manure.

Previous generations had two options when it came to utilizing manure as a fertilizer. It could be applied fresh. While fresh manure is nutrient-rich, it's also rich in bacteria that is harmful to the soil, and there is a possibility that the harmful bacteria will outcompete beneficial bacteria for survival. Fresh manure is also incredibly expensive to ship due to its high water-content.

The second option would be to cure or age the manure for a time. This allows the harmful bacteria to die off. However, as the manure ages, it also loses much of its nutritional value through evaporation or leaching. Manures are also often laden with the cides and antibiotics that kill off elements of the healthy, soil microbiome that we must establish to rebuild our soils. Dried or cured manures are easier to ship, but often not worth shipping.

Over the past decades, our technology in fertilizer production has improved tremendously though. We now have methods to simultaneously remove water content in manure, kill off any harmful bacteria, and stabilize its nutrients in a timely way. And why wouldn't we want to do so? So, many environmental advocates are pushing for the elimination of the livestock industry to solve the manure problem, but how often have we heard a move to eliminate synthetic, inorganic nitrogen fertilizers? One industry is comprised of a multitude of small business owners, and the other of powerful corporations.

If we could replace synthetic, inorganic nitrogen fertilizers with manure-based, biotic fertilizers, how would the earth look different? Oceanic, dead zones would substantially decrease. The air we breathe wouldn't get any worse than what it is now. Our foods would become more nutrient dense and naturally flavorful. Our people would be healthier, happier, more resilient, and mentally and physically stronger. Fewer young couples would experience the pain of infertility, and more elderly patients might live out their final days surrounded by the comforts of their own homes. Dense, fertile, nutrient-rich soil microbiomes could be rebuilt. We could at least begin to stabilize our soil loss and potentially rebuild the billions of tons of farm soils we've lost.

Known as biotic fertilizers, there is a new class of fertilizers that approaches the soil differently. Biotic fertilizers aren't meant to directly feed plants nitrogen, phosphorus, and potassium. They are designed to feed the soil microbiome so that it can, in coordination with the plants, create complex healing fatty acids, proteins, carbohydrates, and amino acids to feed animals and people. They are designed to help support the soil microbiome to perform its natural roles in hunting, gathering, storing, and feeding complex nutrients to plants. They function more like the slower, natural nutrient recycling and creation process that occurs in uncultivated soils. However, they are customized to meet the needs of the soil microbiome they will be applied to. When we focus on feeding the soil microbiome, we understand that a healthy soil microbiome produces its own, plant-available, organic nitrogen in the form plants prefer, proteins and amino acids. Biotic fertilizers operate on the belief that feeding each soil microbiome what it needs will provide soil microlife with the strength and sustenance to grow nutrient dense foods. We are no longer focused on growing crops; we must focus on growing soils. They will grow the crops for us.

Biotic fertilizers also aren't a means of offloading some excess by-product from another industry. They are reestablishing, maximizing, and enhancing the natural plant-animal loop that has existed since the dawn of time. Biotic fertilizers are not merely fixing a broken animal-soil-plant pathway or returning to the *"good, old days"* of agriculture. Biotic fertilizers and a biotic approach utilize all that has been learned and studied from generations of agriculturists, soil scientists, and a vast network of researchers to make a huge, leap forward in agricultural understanding and ability, far past what any previous generation of agriculturists accomplished. It can only be done because of the faith and perseverance of these pioneers in agriculture who went before us.

A dying or dead soil microbiome is like a hospital patient on the brink of death, but with a strong will to live. It needs a great deal of care and tenderness. It also needs the right diagnostics. When the multitude of mystery illnesses began emerging in the 2000's, people turned to the same classical medical institutions they had always relied upon only to learn that available medical tests were insufficient to diagnose their conditions. Out of this tremendous need, a new branch of medicine emerged known as functional medicine. In functional medicine, medical conditions were not treated as isolated problems of individual organs, but as connected events across an intimately, interconnected body. Functional medicine physicians forewent peddling prescription medication band-aids and surgeries to eliminate individual symptoms and sought to heal the root causes of these illnesses.

Along with the emergence of functional medicine, as a discipline, emerged fantastic new diagnostic laboratories with abilities far beyond basic blood chemistry levels. Just as soil scientists were beginning to view the soil as a biological civilization, rather than a mineral medium, these diagnostic labs and functional physicians were beginning to assess bacterial civilizations and fungal empires. A conventional medicine blood test might measure the amount of B12 present in a patient's bloodstream. A functional medicine blood test would measure the blood cell's response and ability to utilize B12. The conventional medicine patient might be told that their B12 levels were fine, and that their fatigue and anxiety were simply from stress. The functional medicine patient might learn that their cellular DNA was struggling to process B12 and that their fatigue and anxiety could be improved simply by taking a different form of B12. Other functional medicine diagnostics could separate out different strains and percentages of bacteria within the digestive tract or measure the integrity of the cell membranes of the digestive tract itself.

The entire shift was two-fold. Firstly, functional medicine viewed the body as more than merely a host of chemical reactions. Functional medicine took a biological view of the body that sought to understand fungal colonies, chronic viruses, and bacterial biofilms. Functional medicine acknowledged that the body's biological responses were often considerably more intricate than had previously been believed. The second shift was to reassess our definition of health. In conventional medicine, an absence of pain, malady, or injury has long been viewed as "healthy," but in functional medicine, health is the presence of strength, energy, and vitality. It's maximizing a person's biological potential. The absence of symptoms or the inability to locate a cause is not the opposite of sickness. The opposite of illness is thriving.

For decades, we've held a similar chemical approach to our soil microbiome, fields, and crops. As in conventional medicine, chemical and mineral testing have been the gold standard for soil testing. Like basic bloodwork, it's not unhelpful information, but these standards were set back when we believed that soil was purely a mineral medium. Today, we know better. We need a similar shift in our approach to the soil. Like the human body, perhaps the first stop for our soils should be improved diagnostics that reveal to us the health of its microbial communities, as well as its chemical status. Before we can create a healing plan, we need to triage the damage.

Labs already exist that can easily and affordably test soils, not merely for chemical elements, but for the cumulative health of the soil microbiome as well as for its various communities. Growers can send

There Might Be Hope

in soil samples to measure Mycorrhizae, nematodes, active bacteria, total bacteria, active fungi, total fungi, and protozoa. Understanding the overall balance of the soil microbiome within a field or lack of balance will empower the grower to feed specific microbial communities with fertilizers targeted for that community or provide starter bacteria or spores to restore the soil microbiome. We must not forget that 700 different types of bacteria are needed simply to germinate a seed. Because of this and because soil microbial communities naturally change and shift, in collaboration with the plant, throughout the life cycle of the plant, it seems logical that we ought to check and test the soil microbiome multiple times during the growing season and the fallowing season as well. These tests are inexpensive, and the payoffs are enormous.

Another essential tool is testing for both the presence and the quantity of pesticides within the soil. Depending upon the types and amounts of pesticides applied and persisting, these pesticides can destroy some soil communities or shift others, such as bacterial communities, away from beneficial bacteria towards the dominance of harmful bacteria. Pesticide testing is considerably more expensive, and due to the immense quantity of pesticide formulas has room for improvement in its accuracy.

The third component of our diagnostics needs to be measuring the plant Brix levels. Brix levels are a simple, inexpensive, and easy measurement of plant health which directly corresponds, not only to vegetable or fruit flavor and nutrient density, but also how attractive or unattractive the plant is to harmful insects. If the soil microbiome is thriving and balanced and pesticide levels are low to non-existent, then Brix levels should be high. When soil microbiomes are healthy and thriving, plants can maximize their nutrient density. When nutrient density is maximized, fruits and vegetables become sweeter and more flavorful and insects lose interest in devouring them.

Healthy and thriving soils are erosion resistant soils. They are also able to capture and hold water. A healthy, soil microbiome is the difference between productive, semi-arid to arid farmland and barren, sand dunes that follow the winds on their whimsies.

What about the sand dunes of Arabia? Can soil microbiomes like these be remediated sufficiently to support the forests? This remains to be seen. In nature, forests usually are built upon grasslands, over time, but healing desertified soil microbiomes to support grasslands or even crops is within reach. A better start to greening the deserts might be saturating strategic areas with microbial teas. Like kombucha for the soil, microbial soil teas are designed for specific soil needs and contain prebiotics and probiotics to help jumpstart the soil to life. We can also introduce fungal spores to initiate Mycorrhizal colonies. Bearing in mind that Mycorrhizae is a husband in need of a wife, it's important for fields to be simultaneously seeded with a cover crop. This will allow the Mycorrhizae to marry into a plant root system, creating a symbiotic relationship that will result in beginning to restock the soil's carbon pantry with humic and fulvic acids and restore the soil's natural, mini-water reservoirs. The sooner we can begin to build these, the less water we will ultimately need to grow crops, and the less erosion we will experience. This has been the age-old problem with worn out farmlands, such as those surrounding Leptis Magna and across the Middle East. Once soils reach a certain level of depletion in the soil microbiome pantry, the microbes that build and maintain the water reservoirs starve to death. Soil infrastructure gets broken down, and the soils no longer absorb rainwater. Rain still falls in the desert, but there is no life in the soil to capture it.

Once the soil microbiome has been stabilized with some basic infrastructure, it will be ready for the more solid and substantial food of biotic fertilizers to further build the soil microbiome's carbon infrastructure, water storage capacity, and nutrient storehouses. Admittedly, our present understanding of the soil microbiome, like our understanding of the ocean is still quite rudimentary. Despite this, such an approach has been demonstrated to transform sand into loam as soil microbes build complex carbon, fatty acid, protein, and water reservoirs, along with soil infrastructure. Loam is, by definition, a soil with a strong granular structure created by the presence of high organic matter. As the soil microbiome comes back to life, sand will deepen to a soft and then rich brown color. This is no overnight process. The speed and success of soil restoration will depend on how accurately we supply the correct quantity and quality of nutrients a particular soil microbiome needs. It also depends upon appropriate watering, sunlight, and temperature. Some elements are out of our control. Existing pesticide and trans-genes residues that may inhibit soil microbiome growth may be the most challenging element to address though. Nevertheless, with enough resources and determination, sand can be turned into healthy, thriving loamy soil where nutrient dense food can be grown and perhaps, with time, even trees.

We have no lack of eroded, pulverized, and depleted farmlands and former farmlands in desperate need of soil microbiome restoration. These are, increasingly, hedged in by cities and surrounded by mountainous, rocky, ice-covered, and otherwise simply unfarmable lands. Time is of the essence. What brought us here is simple. The massive erosion we've witnessed can be simply summed up as a failure to understand soils as biological civilizations rather than as a mineral medium, leading to the death of our farm soils through synthetic, inorganic nitrogen, pesticides, and trans-genes.

The scientific and practical applications of this momentous shift in agriculture are likely the easiest aspects to adopt. Perhaps the most important factor not yet discussed is that the regreening of our farmlands and deserts would require a changing of our hearts and minds. We are going to have to reimagine what an idyllic farm field looks like. We're going to have to say goodbye to broad, insect-free fields with crusty, greige, hard-pan barren furrows. We probably won't be driving past miles and miles of the same crop. The ideal field for soil microbiome health will likely be narrower and/or shorter than we've grown accustomed to in order to provide close access to soil microbial and Mycorrhizae nurseries. Farms will grow a variety of crops, to ensure the nutritional density of each. Monoculture is as stressful to the soil microbiome as drought. Farmers will find new ways to reembrace diversified crop agriculture. Rather an absence of insects, we're going to be looking for a particular balance of insect species, favoring the beneficial. We're going to rely on them to be a key part of our soil diagnostics to tell us what's missing in the soil microbiome. Rather than an absence of weeds, we're going to be monitoring which weeds are appearing and what they might be telling us about the health of our soil microbiome.

Ultimately, we will need less food and fewer supplements because our foods will become more nutrient dense. Crops will be robust, and although no longer toxic to insects, they will be less desirable to them. Our crops will withstand drought and fungal attacks better. With soil at its peak performance, a grower will likely apply less fertilizer because the soil microbiome is functioning as it was designed to. The fertilizer we do apply will be better absorbed because it is in the form soil civilizations can best utilize. Farmers can focus on only replenishing the nutrients required by their soil microbiome at strategic stages in its development.

We will need more young men and women to pursue the study of the soil microbiome and develop new and enhanced diagnostic tools to help us learn more of vast world of soil science that lies yet undiscovered. Most importantly, we must give farmers freedom and support healthy competition between them, something that hasn't happened since 1933, when FDR first reigned in American agriculture under socialism.

In the 1920's, America's farmers were really good at producing food. Really good. The United States had such an abundance of relatively healthy food that the cost of food dropped substantially. Imagine that. The United States was world renowned for its abundance of relatively healthy, inexpensive food. Immigrants arriving to America are reported to have often been shocked by sheer abundance and affordability of food in America.

This relative abundance of food, likely originated from America's worldview at that time. America had predominately been settled by Christians, often fleeing persecution, if not imminent imprisonment and death for their faith. They had a tough road, creating new lives for themselves out of the wilderness. It would be remiss not mention that contrary to current, popular belief, American Christians *"were the first religious group,"* of all the world religions, *"to find slavery, morally intolerable."*[3] Until that point, every culture on every continent approached slavery within its society as a normal part of life. America's early founders believed that God was inherently good and that He had created man in His image. It was this common, core belief that led them to write that *"all men are created equal, endowed by their Creator with certain unalienable rights."* Their approach to American agriculture was shaped in the belief that God had given man both the authority and responsibility to care for the animals and plants of the earth.

They also believed that man was born into sin and that all the wrongs of the world were the result of sin, either our own or someone else's. They believed that both they themselves and anyone who wronged them would ultimately be held accountable by a just and righteous God. They believed that those who did good would ultimately receive eternal reward. Those who didn't, would suffer loss. These concepts shaped and motivated them in how they responded to others. These were common, American beliefs because generations of Americans learned to read using the same book, the Bible.

They also generally believed in tiers of order and governance, including self-governance, family governance, ecclesiastical governance, and civil governance and that each had distinct roles and responsibilities. Each person had a unique, God-given identity and role in life. Because they believed that God rewards each person individually for their hard work and ingenuity, they valued not only hard work and ingenuity, but also personal property and ensuring that each person received the fruits of their own labor. They held, what would be known today, as an *"abundance mindset."* Because they believed that God would provide them with what they needed, they generally believed in giving, and giving freely, to those less fortunate than themselves, primarily through the church.

Obviously, not everyone held these beliefs in America, and there were pockets of Americans whose views differed considerably. For hundreds of years though, these general beliefs were held by the majority of Americans. These were the foundational beliefs that inspired hard work and provided a framework for American farmers to create an abundance such that the world had rarely, if ever, seen before. Ironically, it was these people who generally valued storing up eternal treasures for themselves more than earthly ones that ultimately laid the foundation for one of the wealthiest, per capita, nations the world has ever seen.

In the mid to late 19th century, a number of theorists and philosophers emerged with radically contrasting views to those which had built America. Charles Darwin and Karl Marx were two of these men. Their beliefs were based on man being the central god of his own life and destiny, the captain of his own ship. They believed that

There Might Be Hope

there was no God to judge, no God to reward, no God to avenge. There was no eternal paradise or damnation. The only happiness they believed in was what they could create for themselves, by whatever means necessary.

In the early 1900's, the world had yet to witness the full misery and atrocities of communism in control of a country. Multiple, related branches of socialist, communist, and anarchist parties had openly built their platforms and followings in American politics since the late 1800's. In the 1920's, both communist and socialist parties offered presidential candidates, as did the Farmer-Labor Party, whose name belied its strong socialist and communist roots and affiliations. Political infighting and divisions hindered most of their progress.

Their ultimate goal was complete totalitarian control. Class warfare, land control, and agriculture lay at the heart of their plans. Leaders within these movements often projected the plight of European and Russian laborers on recently immigrated American farmers and laborers who hadn't quite established themselves in their new country yet and were vulnerable to this propaganda. Around the 1930's, as the world was gradually awakening to the real dangers to life and liberty posed by these ideologies, political infighting and internal divisions intensified to the point that the more formal branches of many of these movements, including the Farmer-Labor Party generally dispersed. Somehow, the ideas persisted though and found their ways into the heart of American agriculture.

"...the plain truth is that the root causes of the agricultural crisis were far removed from agriculture."

~ *"The Agricultural Crisis, 1920-23" by R. R. Enfield*

> "Regarding the Great Depression, ... we did it. We're very sorry. ... We won't do it again."
> ~ Ben Bernanke, November 8, 2002, in a speech given at "A Conference to Honor Milton Friedman ... On the Occasion of His 90th Birthday."

Early American revolutionaries were likely the first, known government to finance their war by printing paper money. As with all, untethered money printing, it didn't take long for the money to be, essentially worthless. George Washington is credited with saying that *"a wagonload of Continentals will hardly purchase a wagonload of provisions."*[4] Paper money was meaningless to the local merchants who would only accept precious metal coins. To solve this problem after the war, the new American government took out a loan, bought precious metal coins, and burned their paper money.

Similarly post-WWI governments sought to finance their recoveries by printing off paper money and the natural consequence of this was extreme inflation in the post-WWI, 1920's. The Federal Reserve, which had only been in existence for a scant 7 years at this time, began to raise interest rates to slow inflation. People suddenly couldn't pay their loans. Then, banks began to fail.

Beginning the early 1920's, agricultural commodity prices took a drastic fall with some dropping by more than 50%. The dairy industry bounced back first, being locally produced and not shipped overseas, but grain crop prices, particularly corn, continued to fall. Following the war, European countries had rapidly printed off money to meet the government's needs and left their citizens with paper practically worth nothing. So, falling prices were initially blamed on global grain market's rapidly contracting, but then and now, deflation and speculative manipulation, brought on by the Federal Reserve, has been considered to be the root of falling agricultural prices.

In response, some farmers began to voluntarily fallow fields. Given that the majority of farmers at that time were small, self-sustaining family farms, it's a shame that the excess grain produced in the 1920's could not have been simply stored away for a rainy day or a decade-long drought. Then, in 1930, severe drought hit. The Midwest and Great Plains were affected first. In 1931, the Black Blizzards came. These great walls of parched soils began rolling across the plains and even into the eastern seaboard, smothering New York City and Washington D.C.

By 1932, farmers were making less than 1/3 the income they had made in 1929. Many were forced to foreclose, worsening the crisis. In March of 1933, Franklin Delano Roosevelt was sworn into office, and in May of 1933, the Agricultural Adjustment Act was passed. It was the end of freedom for American farmers.

"The nine most terrifying words in the English language are: I'm from the Government, and I'm here to help."
~Ronald Reagan

Farmers went from being independent entrepreneurs to being welfare recipients, simply by virtue of their vocation, whether they liked it or not. The initial stated goal of the Agricultural Adjustment Act was parity for agricultural prices, that everyone would get paid the same amount for the same commodity. Soon, the goal become parity of income, that all farmers would earn purchasing power equal to the income of a non-farm worker, between the years of 1909 and 1914. These were considered to be the last good years. The first step to parity was for the government to step in and set prices on agricultural commodities. No longer would the supply and demand be allowed to balance themselves, or farmers be able to earn more for producing a superior product. The second step was to control production. Farmers were paid not to farm. They were paid to leave their fields unplanted or worse, were forced to plow their crops under. If a farmer didn't want to participate in this, it was fine, but in addition to receiving government payments, participating farmers were essentially allowed to sell their remaining produce tax-free. Non-participating farmers had to pay taxes on their crop sales. The result was that an estimated 10.4 million acres of cotton were plowed under in 1933. Across America, fields were left barren.

"In 1932, there were 14 dust storms of regional extent; in 1933, thirty-eight; in 1934, twenty-two; in 1935, forty; in 1936, sixty-eight; in 1937, seventy-two; in 1938, sixty-one; in 1939, thirty.... In Amarillo the worst year for storms was 1935, when they lasted a total of 908 hours. Seven times, from January to March, the visibility in Amarillo declined to zero; one of these complete blackouts lasted eleven hours. In another instance a single storm raged for 3½ days."[5]

Overzealous tractor usage and plowing of the 1920's were likely initial contributors to the Dust Bowl of the 1930's, but leaving fields unplanted and without a cover crop, to protect and support the Mycorrhizae and the soil microbiome was a death sentence.

Parity included not only grains such as wheat and corn, but also encompassed livestock. Beginning in the fall of 1933, the government bought and killed 6 million hogs including pregnant sows. Of the 6 million, an estimated 1 million were processed for meat and donated to the Red Cross. At a time when many Americans were homeless and hungry, if not starving, the remaining 5 million hogs were simply destroyed or "tanked" and rendered inedible. For farmers who still had a herd left after the buy-out, hog prices increased a mere 20 to 30%.[6] Cattle weren't off the table either. In Nebraska, nearly half a million head of cattle were shot and buried in pits to earn Agricultural Adjustment Act money.

The outcry against the Agricultural Adjustment Act of 1933 was substantial, and not because it awarded the Secretary of Agriculture $100,000,000[7], today's equivalent of over 2 billion dollars, for administrative expenses. In 1936, the act was struck down by the Supreme Court as unconstitutional because it was essentially coerced, federal redistribution of wealth. The act took money from agricultural processors to give it to agricultural producers.

"A tax, in the general understanding of the term, and as used in the Constitution, signifies an exaction for the support of the government. The word has never been thought to connote the expropriation of money from one group for the benefit of another.... The act invades the reserved rights of the states. It is a statutory plan to regulate and control agricultural production, a matter beyond the powers delegated to the federal government. The tax, the appropriation of the funds raised, and the direction for their disbursement, are but parts of the plan. They are but means to an unconstitutional end.... If the cotton grower elects not to accept the benefits, he will receive less for his crops; those who receive payments will be able to undersell him. The result may well to financial ruin. The coercive purpose and intent of the statute is not obscured by the fact that it has not been perfectly successful. It is pointed out that, because there still remained a minority whom the rental and benefit payments were insufficient to induce to surrender their independence of action, the Congress has gone further, and, in

the Bankhead Cotton Act, used the taxing power in a more directly minatory fashion to compel submission. This progression only serves more fully to expose the coercive purpose of the so-called tax imposed by the present act. It is clear that the Department of Agriculture has properly described the plan as one to keep a noncooperating minority in line. This is coercion by economic pressure.

The asserted power of choice is illusory....A possible result of sustaining the claimed federal power would be that every business group which thought itself underprivileged might demand that a tax be laid on its vendors or vendees, the proceeds to be appropriated to the redress of its deficiency of income. The supposed cases are no more improbable than would the present act have been deemed a few years ago. Until recently no suggestion of the existence of any such power in the federal government has been advanced. The expressions of the framers of the Constitution, the decisions of this court interpreting that instrument and the writings of great commentators will be searched in vain for any suggestion that there exists in the clause under discussion or elsewhere in the Constitution, the authority whereby every provision and every fair, implication from that instrument may be subverted, the independence of the individual states obliterated, and the United States converted into a central government exercising uncontrolled police power in every state of the Union, superseding all local control or regulation of the affairs or concerns of the states. Hamilton himself, the leading advocate of broad interpretation of the power to tax and to appropriate for the general welfare, never suggested that any power granted by the Constitution could be used for the destruction of local self-government in the states..."

~United States v. Butler et. al.

Despite the 1933's American Agricultural Act being overturned, the federal government's foot was in the door when it came to agricultural control, and the redistribution didn't stop. It merely took a less direct route. Price control or support programs, crop insurance, and subsidies, under a multitude of names have come and gone since then. Today, under our modern Farm Bill, the USDA offers *"more than 60 direct and indirect aid programs for farmers"*[8] and pays a select group of farmers around $20-30 billion a year to ensure farmers receive a minimum price for the crops they plant. Some of the aid, such as crop insurance, is virtually impossible to track, leading to cries of cronyism.[9] Most of the aid goes to large farms growing the staple crops of corn, soy, and wheat. Upwards of 90% of these crops are genetically modified, excluding wheat, which may be soon. The majority of them are also intensively, synthetically fertilized, and drenched with a multitude of pesticides. Why is the U.S. government subsidizing the poisoning our people, our soils, our water, and our skies? If genetic modification and its wonder-cides are so phenomenal at increasing crop yields, shouldn't they pay for themselves? Do they really need for the American people to slip them some Monopoly money under the table, or are they ready to stand on their own two feet? If farming practices produce a nutritionally inferior product, destroy our soils, and are financially unsustainable on their own, on what merits do they deserve to be supported?

It is the farm subsidies that the USDA hands out that drives what farmers grow and what Americans eat. The bulk of subsidies are awarded to grow staple grains such as corn, soy, and wheat. Fruits, vegetables, and organics receive relatively little. The problem isn't in the distribution of the subsidies though; The problem is that the government has made themselves to be the self-appointed middleman. Government bureaucracies simply aren't particularly agile, responsive, or unbiased in representing either consumer or farmers needs and interests.

Nor do subsidies ensure food security and a reliable supply of food staples. On the contrary, they actually promote waste and not responsibility. Unlike other American enterprises, farmers receive government payments regardless of whether their crops fail or thrive, and because of the redundancy of programs, farmers can even be, and often are, paid twice for the same lost crop. Moreover, because agricultural subsidies are neither equally nor equitably distributed amongst American farmers, the ripple effect across American agriculture actually harms non-subsidized farmers who may be either dependent upon these staple crops to feed their livestock or growing the same crops without subsidies. In this sense, non-subsidized farmers are having their own wealth redistributed to their neighbors.

On the one hand, Americans are told that farmers must use poisonous chemicals to feed our world population. If we don't use GM crops and rainbow pesticides, people will starve. On the other hand, American farmers are told that low food prices are their own faults because they are simply producing too much food. Hence, they must accept subsidies if they are to stay afloat financially. Which is it?

The American taxpayer is paying farmers to produce food that consumers themselves may not eat, farmed using principles and technology consumers themselves may not support. The real question may be, who are agricultural subsidies benefiting? Since farm subsidies began in 1933, we've lost 70% of our farms, many due to bankruptcy. If subsidies were supposed to ensure farmers with livable prices for their products and ensure a steady supply of food to the market, they clearly aren't doing their job. They are merely eliminating a certain kind of farmer, the smaller ones. Under the farm welfare state, American farms are growing in size, but shrinking in number as they are increasingly consolidated under increasing corporate and conglomerate control. Have the lure of subsidies and a zero-risk business plan been fundamental in this transition? Is having fewer farms under more consolidated control really in the best interest of our national food security, or could it be that factions within the American government simply prefer the control a consolidated food chain offers them, in keeping with Marxist principles?

In the mid 1980's, at the behest of its farmers, New Zealand repealed its subsidies. The next few years marked a challenging transition period for New Zealand farmers as their entrepreneurial mettle and adaptability were tested. They came through like gold. Only 1% of farmers left farming during the transition. Today, their agriculture thrives on freedom. Ours can too.

No one takes better care of their soil and their livestock than a farmer whose home and family are dependent upon them and hopes to pass them on to the next generation. Period. Our best hope for healing our soils, our land, our water, and our people is to unleash the American farmer again. Our lives and those of future generations depend upon young, brilliant, energetic, entrepreneurial farmers who will roll up their sleeves, educate themselves about the soil microbiome, and invest in their own farmlands. We just need the government to let them alone to do so. Easier said than done.

Chapter 10 All Things New

Lying stretched along a footpath, your head rests on the soft, loamy earth as your eyes gaze up at a cloudless, brilliant, blue sky framed by gently swaying, vibrant, green wheat. The sweet aroma of fertile soils and growing crops mixed with a fresh, salty, cool Mediterranean breeze wafts around you.

"A sandwich would be good," you think, as you sit up and reach into your cotton knapsack to retrieve a warm, roast beef and Swiss cheese sandwich on a freshly, toasted baguette along with a bottle of cool, hibiscus tea. Crickets gently chirp as you feast and massage the cool soil with your toes. In the distance, fishing boats bob up and down in the playful waters. Further off, cargo ships brave the Sardis Straits. They are heading to the great metropolis of Rome, loaded down with wheat, beef, sweet fruits, and fresh vegetables to feed hungry tourists. Perhaps, you will visit there next. Maybe you'll even build your next dream home there, but today and most days really, these simple delights are what feeds your heart. Maybe there will be time for Rome, later. As you stand up and shake the stillness out of your muscles, you pause and soak in the warm, sunshine goodness around you.

"Hey!" a stranger behind you shouts. "Sheesh!!!" A strikingly, beautiful creature saunters up the path towards you. It stops and begins softly clapping and swaying its hips to music you cannot hear. "Like TFW! Like I'm totally stanning this. Like, spill the tea!" It makes a heart with its paws. With a dreamy smile, the creature slowly circles around you gazing at the distant waves. The delight of company comes mixed with other sensations, familiar, but not altogether comfortable. The creature turns to you again, staring intently. "Like...who are you?" It suddenly shouts and jerks away, as if startled.

"Melia, it's ok. It's me," you say more to comfort yourself than her.

She shrieks and begins to shake. Her gaze drops to the earth, and she begins to shuffle off into the wheat. You softly sigh. As you stand to your feet, you see them. A passel of other creatures, of all imaginable and some previously unimaginable shapes, sizes, and types is barreling down the footpath towards you both. Their expressions range from sinister to menacing to laughing hysterically to away with the fairies. You move more quickly to gently take your sister by the shoulders and exit her to a different scenario.

Howling winds and sands pelt the window your arm rests against as you rip off your VR goggles and toss them on the gritty floor in front of you. Your recliner slides back upright. You glance at your neighbor, still reclining in her chair with her face softly glowing, immersed, and enthralled in her VR world. Sweat trickles down her face. Another rolling blackout. You gently brush the sweat away from her cheeks. They took Melia away to another floor some time ago, but the girl now beside you is someone's sister.

Mild regret for chunking your VR goggles sets in as a soft, blue glow at the end of the hallway signals that Nurse Alura is returning. They never let her battery run down. The hallway is otherwise dim,

There Might Be Hope

somewhat from the blackout and more so from the sandstorm. As Nurse Alura rolls towards you down the hallway and past a row of vinyl recliners backed against the wall, the light from her features softly illuminates those you share the hallway with. Most are in their 30s and 40s. A few are still in their 20's. Many are slumped over. Some are softly mumbling to themselves. One is soundlessly wailing. A few are motioning with their hands towards VR world. They are all strapped in.

Alura slows to a stop before you and slightly tosses her hair. "How are you doing?" she smiles as she softly drops a dinner tray across your lap. "I am so excited to show this to you! It's your dinner for tonight. Boiled jellyfish topped with a medley of ground termites, crickets, locusts, and ants. You also have a side salad of hydrophonically, grown salad greens."

A sudden downpour announces its arrival on the roof above you. It never sounds that good in VR world. Streaks of ruddy water now run down the floor to ceiling window. Blood rain. No one sees it but you.

Alura's smile intensifies. "How *are* you?" she repeats herself. Her eyes narrow as she turns toward the goggles on the floor and slowly rolls to retrieve them. Her icy, silicone fingers slide them back over your head. Her eye sensors pierce through the goggles locking onto your eyes. Her smile hasn't changed. "Did your little world get overrun by strangers again? Oh, that's just *so* terribly sad...Is there anything I can get for you?"

"I'm fine," you say, being conscious to maintain relaxed breathing.

"*Wonderful,*" she glares. She turns and slowly rolls back down the hallway. Looking back out the window, the short-lived, rain band has passed and streaks of lightning now mingle with the angry sands that are thrashing the windows once more.

You remember real farms. Your grandfather had one you and your sister played on as kids. It wasn't that long ago.

They say on the dark web that real farms do still exist. If they did exist though, only octillionaires could afford such things now. Looking down at your gaunt arms and bloated belly, you remember the taste of beef and cheese as a kid. Even if you had the money, your social credit score isn't that good. The VR goggles on the floor will set you back another 3 points…unless the fines have gone up again. It may be up to 5 points now.

Tiny, ant heads stare up at you from your plate. You're not touching it. You're not about to have another adverse reaction this evening. You'd rather starve to death.

You return to your window view. The winds have lessened and the faint, semi-circle of a setting sun has appeared across the old, dilapidated parking lot. Thankfulness fills your heart. People with better social credit scores don't have windows anymore, not real ones anyway. They fill their virtual windows with the kinds of virtual paradises and habitats you used to create at work each day. Once your short-term memory started to lose its edge though, your productivity slid, and the better gigs started going to other creators.

You do, vividly, remember the insects taking over American farmlands and cities. They devoured nearly everything in sight and swarmed around the doorsteps. People had no choice but to flock indoors for relief. Then, the sandstorms came. Dirt and insects swarmed and crawled, their way into crevices homeowners didn't even know existed. It was easier to convert old office spaces into makeshift housing and hospitals than to seal off single, family homes from invasion. Besides, no one was having families anymore anyways and caring for multiple dementia patients within one family at home simply wasn't realistic. Thankfully, savvy investors had been prepared. If they hadn't have been, who would have survived? They converted former, online shopping warehouses into vertical garden spaces illuminated with glowing LEDs. They used all the latest science, or at least the science that most benefited their pocketbooks.

Living in a tech dystopia is actually our least likely future scenario. From this vantage point, we really only have two realistic options. One is healing our soil microbiome, and the other is complete global catastrophe and the obliteration of civilizations.

The organic movement that followed the onset of commercial GMO farming roughly 20 years ago, is actually on the decline, both in America and Europe. The number of farms transitioning to become organically certified in America has declined 71% since 2008.[1] To certify an organic field, farmers must prove that the field hasn't been sprayed with any unapproved substances during the last 3 years. This can be an immensely difficult financial transition, and given the persistence of pesticides within the soil, under this rule, there are probably plenty of newly certified organic fields that likely are and will continue to pass on inorganic pesticides to consumers for years to come. So, this might not be our most reliable benchmark.

In 1933, the first contract between a poultry grower and a large, poultry feed company was signed.[2] By 1955, only 10% of poultry production was independently owned. The other 90% was vertically integrated. In vertical integration, the parent company owns everything---the chickens, the feeds, the antibiotics---sometimes even the chicken houses. Poultry farmers provide the land and labor. The swine industry became similarly, vertically integrated shortly thereafter. Farmers made less money than they had made operating independently, but by essentially becoming a contract employee, they also lowered their inherent financial risk.

The Rhode Island Red, the Plymouth Rock, the Wyandotte, the Buckeye, and the Jersey Giant were but a few of the chicken breeds raised by American farmers prior to vertical integration. The breeds had been carefully selected for desirable genetic traits that suited the needs of a particular farmer and gained popularity with other farmers with similar needs and farming practices. A farmer in Minnesota and a farmer in Alabama needed livestock that could thrive in their respective environments, with their unique climates and predators. Vertically integrated companies, such as Tyson chicken, realized that by growing chickens indoors, they could control the environment and thus focus on only the genetic traits that grew as much meat as possible as quickly as possible.

Vertical integration enabled agricultural feed companies to grow into corporations. Their success next inspired the vertical integration of the swine industry which began in the 1960's. The final frontier was the beef industry. In the early 2000's, an attempt was made by a group of college professors to create the first vertically integrated beef operation. It failed miserably, even laughably. No further attempts seem to have been made after this. Unlike chickens and pigs, raising cattle indoors from birth to slaughter simply doesn't work and because of this, the cattle industry is a wild one that the powers that be in agriculture today cannot foreseeably control or dominate. If the public can be convinced to give up beef of their volition, under the guise of environmentalism, they won't need to. If the general public can be persuaded that cattle are bad for the environment or bad for their health, neither of which is true, then the goal of complete agricultural production control by a few limited entities is in clear sight. Cattle aren't patentable. Genes are. Seeds are. Pesticides are. Fertilizers can be. An insatiable ambition for control seems to have overtaken the powers that be in American agriculture.

We didn't need modern chemicals and genetic engineering to destroy our soils; Cultures across the world and throughout human history have destroyed their soils simply by neglecting proper nutrition and rest. The reality is that, despite our incredible advances

There Might Be Hope

in agricultural knowledge and ability, our understanding of the microbial kingdoms within our soils, plants, animals, and ourselves is still very poor and very small. We've mastered sailing around the bay and believe we're ready to dominate the high seas. Our arrogance might be the death of us. It may be too late. The amount of soil we've lost is historically irreversible within our lifetime. As they paint our sunsets red and gold, barren lands of past civilizations are messengers warning us of impending danger, if not doom. Our remaining soils are saturated with halogen-based pesticides, arsenic, and lead. Genetically modified, trans-soil microbes are being flirted with in labs across the world by scientists and corporations with either a death wish or more money than sense. Bioaccumulation is suffocating our oceans and fumigating our skies. Is it really necessary to kill our environment to feed ourselves? This perhaps is one of the biggest lies we've bought into over the past century. The ends do not justify the means. We aren't succeeding in saving our crops from insects, but we may well be exterminating ourselves trying.

In the long-run, who did an over-abundance of cheap food really benefit? The farmer? The consumer? The agri-corporation? Was the push to adopt chemical agriculture really about feeding a starving world or controlling it? Perhaps, what began with good intentions was sabotaged along the way. These things happen.

"There may yet be hope." ~ Lamentations 3:29

Leptis Magna tomorrow doesn't have to be the Leptis Magna it is today. The technology exists to transform Leptis Magna from a UNESCO site semi-encircled with an encroaching desert to a UNESCO site semi-encircled with rich fields of wheat and lush pastures. The technology exists to make the people of Libya food-independent today, as they once were 2 millennia ago. In 300 A.D., farmers may have understood the problem, but they didn't have an effective solution. Today, we have both. We understand that insects plague crops that are nutritionally deficient and that by addressing imbalances within the soil microbiome, we can leverage the microbiome to feed those crops and restore missing nutrients to them that will make them unpalatable to insects. We know that if we can feed the soil microbiome to a balanced and healthy state, we can protect it from erosion. We can even utilize the power of soil fungi and bacteria to transform barren sand into thriving soil civilizations that can once more support plant life to feed animals and people. Is biotic farming financially sustainable? A brief glance at the nutritional supplements industry says it can be if we empower the free-market to work and let the consumer drive production.

The county fair was once an integral part of American life. Local farmers brought their best to be judged and evaluated. Winners received approbation and their livestock and produce were acclaimed and sought out. Today, outside of the farmer's market, individual farmers are scarcely recognized or rewarded for producing food of exceptional taste or nutrient density. They are rewarded for producing food of exceptional beauty and conformity in that beauty. Produce and grain are generally pooled with other produce or grains from many other farms. This enables a consistent and abundant food supply to grocery stores, but it also means that today, we farm to the middle. Farmers want their produce and grain to meet a standard of physical beauty and be uniform in appearance with what is grown on other farms. For lack of a means of easily measuring nutrient content, Americans have long relied on the physical perfection of their foods to reflect their food's flavor and nutrition. In response to America's preoccupation with appearance, producers have responded by creating

There Might Be Hope

designer produce with the appearance of perfection. Producers have a double incentive to create foods with the appearance of perfection. Insta-worthy produce not only sells better with consumers, but the genes are also patentable. What incentive have they had for focusing on produce nutrient content?

Nutrient content translates to flavor. Perfect appearance does not necessarily imply either. Neither does the organic label. It merely guarantees that the food is less toxic than conventionally grown food. Differences in organic soil management can result in vast differences in nutrient content between organic products. At times, it also can result in minimal differences in nutrient content with conventional, pretty produce. An organic farmer who takes meticulous care of his soil microbiome will inevitably produce food of higher nutrient density and better flavor. At the grocery store though, his product is put on the same shelf as the organic farmer who is simply farming without pesticides. The latter's produce will not have as high of nutritional value or be as flavorful, but both are presented to the public as "organic" leaving the general public to be rightfully skeptical of spending extra for a product that looks no different and doesn't consistently taste as good or better than a less expensive, conventional product. Rather than the sole measure of produce being "organic" or "conventional", we need to give the consumer the ability to compare the general nutrient content between produce, and we need to encourage healthy competition between farmers. If a farmer is really talented at producing a particular crop that is above average in nutrient value and flavor, shouldn't they be rewarded for that?

We can easily and inexpensively test the nutrient density of a fruit or vegetable using a Brix refractometer. Labeling produce with their Brix scores would enable the consumer to easily chose the most flavorful and nutritious produce. Grocery stores could easily and inexpensively measure this themselves. What if we simply gave people the transparency they deserve to make informed food choices today and let consumers drive the market? Food conglomerates would likely hate this and fight to prevent it, but those who took the initiative to test and advertise the nutrient density of their crops would likely be rewarded for it.

No mandates. No farmers or producers driven out of business by increasingly burdensome government regulations. Let's have a simple fair fight between producers to do their homework, apply the latest soil knowledge to their crops, and let the most nutrient dense, poison-free products win. Farmers would be incentivized to use for the practices and amendments that grow the most flavorful, nutrient dense food possible. Instead of a world food prize for who can produce the most food, let's have a food prize for farms that produce the most nutrient-dense foods. Let's cast-off the American farm welfare state and pursue a course of American agricultural excellence.

What if the American people weren't nutritionally starved and chemically manipulated? What would our world look like if our crops had not just basic elements like protein, fat, and sugar, but the complex proteins, fatty acids, carbohydrates, and minerals that they held 150 years ago? What if Americans no longer had to rely on supplements for their nutrition? What if American obesity became a by-gone, historic-era, not a living, national endemic? What if streams, rivers, and oceans were no longer dominated by cides and erosion? What if the fish and seafood you ate wasn't loaded with the pesticides that accumulated down the food chain?

No more synthetic, inorganic nitrogen. No more pesticide cocktails. No more needless CAFO waste. No more trans-genes and genetically modified foods. Such a world feels impossible, but like payphones, box screen t.v.s, film cameras, c.d. players, and floppy disks---all technologies eventually run their course. It's time for agriculture to move forward. We don't have half a century of farmable soil left, and even if we did, we might not have a population left to feed by the end of those 50 years.

Of late, global leaders have wielded environmental issues as a banner to force or guilt citizens into making financially crippling life changes and consolidate their own power, but top-down government programs have historically been universally ineffective and wasteful, regardless of who is in charge. The most historically life-giving and abundance promoting changes have always come from freedom. Maximum change comes from giving people choice and empowering them on the individual level. If our soils and the people who live on our planet are to have a chance, they must have a choice.

If our farmers have lacked freedom and our consumers have lacked choice, it is possible that one group may have been given too free a rein of both.

"Your scientists were so preoccupied with whether or not they could, they didn't stop to think if they should." ~ Jurassic Park

We began the 20th century with a battle between chemical versus biological agriculture. We can clearly now see the superiority of biological agriculture. Perhaps, the next emerging question is an even more difficult one. In the past, almost 40 years, agricorporations have taken to patenting everything that for millenia, humankind had taken for granted as unpatentable, from seeds to genes, the foundations of all that we physically are. It seems intuitive that agriculture companies with so much yet to discover about the soil microbiome, would focus their efforts on maximizing its health and vitality. However, companies are scrambling in laboratories, not to study the microbiome and maximize its rarely realized full potential, but to replace it with newly patented, genetically-engineered soil microbes. While there is no lack of need and plenty of wealth to be made in rebuilding our farm soils, the immediate profit of creating a patented product that can supplant nature is far more immense and, evidently, tempting.

Are we simply expanding an ideology which has already proven to be catastrophic for human health from above the soil to below the soil? We don't yet know a fraction of the life that exists below or has existed below the surface of the soil. How can we boldly introduce more trans-species, again, that will unleash, likely global, unknown effects, merely to enhance someone's financial portfolio....again?

The essence of our trouble isn't really the soil. The essence of our trouble isn't a particular farming practice or the lack thereof. It isn't government policy or the lack thereof. It isn't even knowledge or the lack thereof. The essence of our trouble is the heart of man. Whose ideas about nature do we believe are superior, God's ideas or ours? Will we seek out God's design and plan for His creation or insist upon our own? However, mankind answers these questions will determine our future, and yet, regardless of how we answer these questions, one thing is certain.

"Behold, I make all things new." ~ Revelation 21:5

NOTES

1. DUST TO DUST

1 Rowlett, Russ, "Lighthouses of Libya", November 5th, 2012., The University of North Carolina at Chapel Hill, www.unc.edu/~rowlett/lighthouse/lby.htm
2 Munzi, Massimiliano et al. "A Topographic Research Sample in the Territory of Lepcis Magna: Sīlīn." Libyan Studies 35 (2004): 11–66. Web.
3 Pucci, Stefano & Pantosti, D. & De Martini, Paolo Marco & Smedile, Alessandra & Cirelli, Enrico & Pentiricci, Massimo & Musso, Luisa. (2010). "Environment-man relationships in historical times: the balance between urban development and natural forces at Leptis Magna (Libya)." 7/18/13, mc.manuscriptcentral.com/Holcene
4 Kron, Geoffrey, "Roman Agriculture" University of Victoria, p. 3,4, https://www.academia.edu/436589/Roman_Agriculture
5 Munzi, Massimiliano et al. "A Topographic Research Sample in the Territory of Lepcis Magna: Sīlīn."
6 USDA ERS. "U.S. Fruit and Vegetable Import Value Outpaces Volume Growth". July 13, 2022, https://www.ers.usda.gov/data-products/chart-gallery/gallery/chart-detail/?chartId=104212.
7 "Food Miles" NRDC, Nov. 2007 https://foodhub.org/files/resources/FoodMiles.pdf ,
8 Afribiz: Making Business Happen in Africa, "Libya: Economic Developments 2013", 8/16/2013, http://www.afribiz.info/content/libya-economic-developments-2013

9 Munzi, Massimiliano et al. "A Topographic Research Sample in the Territory of Lepcis Magna: Sīlīn."
10 Lowdermilk, Dr. Walter Clay, "Conquest of the Land Through 7,000 Years" August 1953, USDA NRCS Agriculture Information Bulletin No. 99
11 Ibid
12 Essington, Dr. Michael E., Class Notes, Soil Chemistry, University of Tennessee, Web.utk.edu/~drtd0c/SoilErosion.pdf
13. https://www.nrcs.usda.gov/Internet/FSE_MEDIA/stelprdb1041883.png
14 NASA Earth Observatory, "Mississippi River Sediment Plume" https://earthobservatory.nasa.gov/images/1257/mississippi-river-sediment-plume
15 Arnon, Robert, "Amazing soil loss in Great Plains Region," The Western Producer, Jan. 19th. 2017 https://www.producer.com/2017/01/survey-reveals-amazing-soil-loss-in-great-plains-region
16 Nachtergaele, Freddy, FAO 2015 "Status of the WSR Main Report" https://www.fao.org/3/i5199e/i5199e.pdf
17 Rao Enming Xiao Yi Ouyang Zhiyun Yu Xinxiao "National assessment of soil erosion and its spatial patterns in China." Ecosyst Health Sustain. 2015; 1(4):1-10. DOI:10.1890/EHS14-0011.1
18 USDA, FSA, "Conservation Reserve Program Fact Sheet", 2022, https://www.fsa.usda.gov/Assets/USDA-FSA-Public/usdafiles/FactSheets/2019/conservation-reserve_program-fact_sheet.pdf
19 "Acute Food Insecurity" https://www.fews.net/ , January 10, 2017, USAID
20 Ganders, Dana, "Wasted: How America Is Losing Up to 40% Of Its Food From Farm To Fork To Landfill" NRDC, Aug. 2012, https://www.nrdc.org/sites/default/files/wasted-food-IP.pdf.
21 Gro Intelligence, "World Wheat Reserves Outside of China to Drop to 14-Year Low", 22 August 2022, https://www.gro-intelligence.com/insights/world-wheat-reserves-to-drop-to-14-year-low

2. FAITH UNTIL SIGHT

1 Waksman, Selman A. "Soil Microbiology",1952, New York, Wiley, https://archive.org/details/soilmicrobiology00waks
2 Ibid page 62
3 Chabbousou, Francis,"Healthy Crops, A New Agricultural Revolution", 2004, Jon Carpenter Publishing
4 Kennedy, A.C. and Smith, K.L., "Soil Microbial Diversity And the Sustainability of Ag Soils" , 1995, USDA-ARS 1995 DOI: 10.1007/978-84-011-0479-1-6
5 Amaranthus, Mike and Simpson, Larry, Acres USA, April 2011, Vol. 41 p 4
6 Elliott, A.L. and Davis, J.G., "Phosphorus Fertilizers For Organic Farming Systems-0.569" Colorado State Extension Service https://extension.colostate.edu/topic-areas/agriculture/phosphorus-fertilizers-for-organic-farming-systems-0.569/
7 Baylis, G.,(2006). "Effect of vesicular-arbuscular mycorrhizas on growth of Griselinia littoralis (Cornaceae)". New Phytologist. 58. 274 - 278. 10.1111/j.1469-8137.1959.tb05358.x.
8 Kothari, S.K., Marschner, H., and Romheld, V., (1990), "Direct and indirect effects of VA mycorrhizal fungi and rhizosphere microorganisms on acquisition of mineral nutrients by maize (Zea mays L.) in a calcareous soil." New Phytologist, 116: 637-645. https://doi.org/10.1111/j.1469-8137.1990.tb00549.x
9 Baylis , G.T.S., (1967), "Experiments on the Ecological Significance of Phycomycetous Mycorrhizas". New Phytologist, 66: 231-243. https://doi.org/10.1111/j.1469-8137.1967.tb06001.x.
10 Aloui, A., Recorbet, G., Gollotte, A., Robert, F., Valot, B., Gianinazzi-Pearson, V., Aschi-Smiti, S. and Dumas-Gaudot, E. (2009), "On the mechanisms of cadmium stress alleviation in Medicago truncatula by arbuscular mycorrhizal symbiosis: A root proteomic study." Proteomics, 9: 420-433. https://doi.org/10.1002/pmic.200800336
11 Amaranthus, Mike, and Simpson, Larry, Mycorrhizae.com/faqs/activities-mycorrhizal-fungi 3/7/17 3:48pm

12 Ibid
13 Alrajhei, K., Saleh, I., & Abu-Dieyeh, M. H. (2022). Biodiversity of arbuscular mycorrhizal fungi in plant roots and rhizosphere soil from different arid land environment of Qatar. Plant Direct, 6(1), e369. https://doi.org/10.1002/pld3.369
14 Hartmann, Martin et al. "Distinct Soil Microbial Diversity Under Long-Term Organic and Conventional Farming" The ISME Journal 9.5 (2015): 1177-1194. PMC. Web. 21 Sept. 2018
15 Gdanetz, Kristi, et al. "The Wheat Microbiome Under Four Management Strategies, and Potential for Endophytes in Disease Protection" 2017, Phytobiomes Journal Volume 1, Number 3 Pages 158-168 https://doi.org/10.1094/PBIOMES-05-17-0023-R
16 Trap, J., Bonkowski, M., Plassard, C. et al. "Ecological importance of soil bacterivores for ecosystem functions." Plant Soil 398, 1–24 (2016). https://doi.org/10.1007/s11104-015-2671-6
17 Howe AC, Jansson JK, Malfatti SA, Tringe SG, Tiedje JM, Brown CT. "Tackling soil diversity with the assembly of large, complex metagenomes." Proc Natl Acad Sci U S A. 2014 Apr 1;111(13):4904-9. doi: 10.1073/pnas.1402564111. Epub 2014 Mar 14. Erratum in: Proc Natl Acad Sci U S A. 2014 Apr 22;111(16):6115. PMID: 24632729; PMCID: PMC3977251.
18 "Soil Microrganism-Actinomycetes" https://www.agriinfo.in/?page=topic&superid=5&topicid=148
19 Adegboye, Mobolaji & Babalola, Olubukola. (2012). "Taxonomy and ecology of antibiotic producing Actinomycetes." African journal of agricultural research. 7. 2255-2261. 10.5897/AJARX11.071.
20 Polpaa, A.J., Bhavanath, J., "New Dimensions of Research on Actinomycetes: Quest for Next Generation Antibiotics" Frontiers in Microbiology (19 AUG 2016) DOI: 10.3389/fmicb.2016.01295"
21 Waksman, Selman A. , "Soil Microbiology" John Wiley and Sons, 1952, pg. 77

22 Bentley, S., Chater, K., Cerdeño-Tárraga, AM. et al. "Complete genome sequence of the model actinomycete Streptomyces coelicolor A3(2)." Nature 417, 141–147 (2002). https://doi.org/10.1038/417141a

23 Kleber, Marcus Lehmann, Johannes, "The Contentious Nature of SOM—What is it and why should we care what it is?" Oregon State University https://users.unimi.it/ricicla/Lezioni/convegno_18-06-15/2-Kleber.pdf

24 Trap, J., Bonkowski, M., Plassard, C. et al. "Ecological importance of soil bacterivores for ecosystem functions." Plant Soil 398, 1–24 (2016). https://doi.org/10.1007/s11104-015-2671-6

25 Pausch, J., et al., "Fluxes of root-derived carbon into the nematode micro-food web of an arable soil," Food Webs (2016) http://dx.doi.org/10.1016/j.fooweb.2016.05.001

26 Peglar, Tori, "1995 Reintroduction of Wolves in Yellowstone" May 13, 2022 https://www.yellowstonepark.com/park/yellowstone-wolves-reintroduction

3. NATIVE FARMERS

1 Sjoerd Willem Duiker, Ph.D., CCA "Avoiding Soil Compaction" PennState Extension, February 12, 2005, https://extension.psu.edu/avoiding-soil-compaction

2 Pryor, F. L. (1985). The Invention of the Plow. Comparative Studies in Society and History, 27(4), 727–743. http://www.jstor.org/stable/178600

3 Amaranthus, Mike, and Simpson, Larry, Mycorrhizae.com/faqs/activities-mycorrhizal-fungi 3/7/17 3:48pm

4 Mulvaney RL, Khan SA, Ellsworth TR. "Synthetic nitrogen fertilizers deplete soil nitrogen: a global dilemma for sustainable cereal production." J Environ Qual. 2009 Oct 29;38(6):2295-314. doi: 10.2134/jeq2008.0527. PMID: 19875786.

5 "World Fertilizer Trends and Outlook 2022" FAO Rome, 2019, https://www.fao.org/3/ca6746en/ca6746en.pdf

6 Maier RJ, Moshiri F. "Role of the Azotobacter vinelandii nitrogenase-protective shethna protein in preventing oxygen-mediated cell death." J Bacteriol. 2000 Jul;182(13):3854-7. doi: 10.1128/JB.182.13.3854-3857.2000. PMID: 10851006; PMCID: PMC94562.
7 Mooshammer, M., Wanek, W., Hämmerle, I. et al. "Adjustment of microbial nitrogen use efficiency to carbon:nitrogen imbalances regulates soil nitrogen cycling." Nat Commun 5, 3694 (2014). https://doi.org/10.1038/ncomms4694
8 Schimel, Joshua & Bennett, Jennnifer. (2004). "Nitrogen Mineralization: Challenges of a Changing Paradigm." Ecology. 85. 591–602. 10.1890/03-8002.
9 Johnson NC. "Can Fertilization of Soil Select Less Mutualistic Mycorrhizae?" Ecol Appl. 1993 Nov;3(4):749-757. doi: 10.2307/1942106. PMID: 27759303.1-5
10 Canfield DE, Glazer AN, Falkowski PG. "The evolution and future of Earth's nitrogen cycle." Science. 2010 Oct 8;330(6001):192-6. doi: 10.1126/science.1186120. PMID: 20929768.
11 Ibid
12 Juge C, Samson J, Bastien C, Vierheilig H, Coughlan A, Piché Y. "Breaking dormancy is spores of the arbuscular mycorrhizal fungus Glomus intraradices: a critical cold-storage period." Mycorrhiza. 2002 Feb;12(1):37-42. doi: 10.1007/s00572-001-0151-8. PMID: 11968945.

4. *Fast Kill, Slow Kill*

1 Klein, Christopher. "10 Things You May Not Have Known About the Dust Bowl", https://www.history.com/news/10-things-you-may-not-know-about-the-dust-bowl Mar. 21st, 2019
2 Wheeler, Charles M. Lt. Col, "Control of Typhus in Italy 1943-1944 by Use of DDT", American Journal of Public Health. Vol. 36. https://ajph.aphapublications.org/doi/pdf/10.2105/AJPH.36.2.119

3 Tsutsui, William M., "Looking Straight at 'Them!' Understanding the Big Bug Movies of the 1950s". Environmental History. Volume 12, issue 2, pages 237-253. DOI: https://dx.doi.org/10.1093/envhis/12.2.237.

4 EPA. "DDT Regulatory History: A Brief History (to 1975)" https://archive.epa.gov/epa/aboutepa/ddt-regulatory-history-brief-survey-1975.html

5 Ibrahim MA, Griko N, Junker M, Bulla LA. "Bacillus thuringiensis: a genomics and proteomics perspective." Bioeng Bugs. 2010 Jan-Feb;1(1):31-50. doi: 10.4161/bbug.1.1.10519. PMID: 21327125; PMCID: PMC3035146.

6 Kado, Clarence I., "Historical account on gaining insights on the mechanism of crown gall tumorigenesis induced by Agrobacterium tumefaciens" Frontiers in Microbiology. Aug. 7th, 2014. doi: 10.3389/fmicb.2014.00340 https://www.ncbi.nlm.nih.gov/pmc/articles/PMC4124706/

7 Sparks, Thomas C. et. al., "IRAC: Mode of Action Classification and Insecticide Resistance Management" Pesticide Biochemistry and Physiology, Vol. 121, June 2015, Pgs. 122-128. https://doi/10.1016/j/pestbp.2014.11.014

5. AN UNLOVED FLOWER

1 Petersen, Gale E. "The Discovery and Development of 2, 4-D", Agricultural History, Vol. 41, No. 3 (Jul., 1967), pp. 243-254 (12 pages)

2 "2,4-D RED Facts" EPA-738-F-05-002 June 30, 2005, https://www3.epa.gov/pesticides/chem_search/reg_actions/reregistration/fs_PC-030001_30-Jun-05.pdf

3 Troyer, J.R. 2001. "In the beginning: the multiple discovery of the first hormone herbicides." Weed Sci. 49:290-297.

4 Petersen, Gale E. "The Discovery and Development of 2, 4-D", Agricultural History, Vol. 41, No. 3 (Jul., 1967), pp. 243-254 (12 pages)

5 Timmons, F. (1970). "A History of Weed Control in the United States and Canada." Weed Science,18(2), 294-307. doi:10.1017/S0043174500079807

6 Schonnbrunn, Ernest et al. "Interaction of the Herbicide Glyphosate with Its Target Enzyme 5-Enolpyruvylshikimate 3-Phosphate Synthase in Atomic Detail." Proc. Natl. Acad. Sci. USA. 2001 Feb. 13; 98 (4): 1376-1380. 10.1073/pnas.98.4.1376

7 Arney, Kat. "Shikimic Acid", https://www.chemistryworld.com/podcasts/shikimic-acid/1010356.article June 8, 2016 10/16/18

8 Cox C, Surgan M. "Unidentified inert ingredients in pesticides: implications for human and environmental health." Environ Health Perspect. 2006 Dec;114(12):1803-6. doi: 10.1289/ehp.9374. PMID: 17185266; PMCID: PMC1764160.

9 N. Defarge, J. Spiroux de Vendômois, G.E. Séralini, "Toxicity of formulants and heavy metals in glyphosate-based herbicides and other pesticides" Toxicology Reports, Vol. 5,2018, p. 156-163 https://doi.org/10.1016/j.toxrep.2017.12.025.

10 Harris, Paul, "Monsanto sued small farmers to protect seed patents – report", The Guardian, 12 Feb 2013, https://www.theguardian.com/environment/2013/feb/12/monsanto-sues-farmers-seed-patents

1111 Gillam, Carey, "Organic growers lose decision in suit versus Monsanto over seeds" Reuters, June 10, 2013, https://www.reuters.com/article/us-monsanto-organic-lawsuit/organic-growers-lose-decision-in-suit-versus-monsanto-over-seeds-idUSBRE9590ZD20130610

12 Francis Chaboussou. "Healthy Crops, A New Agricultural Revolution". Jon Carpenter Publishing, Alder House. 2010.p.7

13 Ibid p.31

14.Ibid

6. THE DIVERSIFIED FARMER

1 "Union Stockyards, Part 2: Technology That Changed Chicago," Chicago Public Library, January 21, 2014 https://www.chipublib.org/blogs/post/technology-that-changed-chicago-union-stockyards-part-two/

2 "GMO Crops, Animal Food, and Beyond" USDA, 08/03/2022 https://www.fda.gov/food/agricultural-biotechnology/gmo-crops-animal-food-and-beyond

3 Robinson, Ashley, "Cancer-causing cattle disease increasing across Canada," https://www.producer.com/news/cancer-causing-cattle-disease-increasing-across-canada-2/ Published: November 26, 2015

4 Shruthi PJ, Sujatha K, Srilatha CH, et al. "Incidence of different tumors in bovines." Open Access J Sci. 2018;2(4):220-222. DOI: 10.15406/oajs.2018.02.00076

5 Casida, John E. "Pest Toxicology: The Primary Mechanisms of Pesticide Action" Environmental Chemistry and Toxicology Laboratory, Dept. of Environmental Science, Policy & Mgmt., University of California. Chem. Res. Toxicol. 2009, 22 (4) pp. 609-619, DOI:10.1021/tx8004949

6 Sparks, Thomas C. et al, "IRAC: Mode of Action Classification and Insecticide Resistance Management" Pesticide Biochemistry and Physiology, Vol. 121, June 2015, Pgs. 122-128 https://doi/10.1016/j.pestbp.2014.11.014

7 Bueno de Mesquita, Clifton P., et al "Adverse impacts of Roundup on soil bacteria, soil chemistry and mycorrhiza fungi during restoration of a Colorado grassland" Applied Soil Ecology Volume 185, May 2023, 104778 https://doi.org/10.1016/j.apsoil.2022.104778

8 Erickson PS, Kalscheur KF. "Nutrition and feeding of dairy cattle." Animal Agriculture. 2020:157–80. doi: 10.1016/B978-0-12-817052-6.00009-4. Epub 2020 Jan 24. PMCID:

9 "Lecture: Inhibition of Photosynthesis Inhibition at Photosystem II" https://www2.lsuagcenter.com/weedscience/pdf/AGRO4070/Handout11.pdf, 4/27/2023

10 "Protoporphyrinogen Oxidase (PPO) Inhibitors" University of California, Division of Agricultural and Natural Resources. https://herbicidesymptoms.ipm.ucanr.edu/MOA/PPO_inhibitors/ 4/27/2023

11 DeFelice, Michael, "PPO Inhibitor (Cell Membrane Disruptor) Herbicides" https://www.pioneer.com/us/agronomy/ppo-inhibitor-herbicides.html 4/27/2023

12 van Almsick, Andreas. "New HPPD-Inhibitors – A Proven Mode of Action as a New Hope to Solve Current Weed Problems" Outlooks on Pest Management, Volume 20, Number 1, February 2009, pp. 27-30(4)

13 Welsh, J., Braun, H., Brown, N., Um, C., Ehret, K., Figueroa, J., & Boyd Barr, D. (2019). "Production-related contaminants (pesticides, antibiotics and hormones) in organic and conventionally produced milk samples sold in the USA." Public Health Nutrition, 22(16), 2972-2980. doi:10.1017/S136898001900106X

14 Zhang L, Jia Q, Liao G, Qian Y, Qiu J. Multi-Residue Determination of 244 "Chemical Contaminants in Chicken Eggs by Liquid Chromatography-Tandem Mass Spectrometry after Effective Lipid Clean-Up." Agriculture. 2022; 12(6):869. https://doi.org/10.3390/agriculture12060869

15 Aulakh, R., Gill, J., Bedi, J., Sharma, J., Joia, B., & Ockerman, H. (0000). "Organochlorine pesticide residues in poultry feed, chicken muscle and eggs at a poultry farm in Punjab, India." Journal of the science of food and agriculture, 86, 741-744. doi: 10.1002/jsfa.2407

16 Shaner, D.L. (2004), "Herbicide safety relative to common targets in plants and mammals." Pest. Manag. Sci., 60: 17-24. https://doi.org/10.1002/ps.782

17 Pritchard J. Alberta. "Organophosphate toxicity in dairy cattle." Can Vet J. 1989 Feb;30(2):179. PMID: 17423244; PMCID: PMC1681043.

18 Prakash N, Narayana K, Murthy GS, Moudgal NR, Honnegowda. "The effect of malathion, an organophosphate, on the plasma FSH, 17 beta-estradiol and progesterone concentrations and acetylcholinesterase activity and conception in dairy cattle." Vet Hum Toxicol. 1992 Apr;34(2):116-9. PMID: 1509669.

19 Kim J-E, Lee H-G. "Amino Acids Supplementation for the Milk and Milk Protein Production of Dairy Cows." Animals. 2021; 11(7):2118. https://doi.org/10.3390/ani11072118

20 Burkholder J, Libra B, Weyer P, Heathcote S, Kolpin D, Thorne PS, Wichman M. "Impacts of waste from concentrated animal feeding operations on water quality." Environ Health Perspect.

2007 Feb;115(2):308-12. doi: 10.1289/ehp.8839. Epub 2006 Nov 14. PMID: 17384784; PMCID: PMC1817674.

7. DEAD ZONES

1 Kontogiannatos, Dimitrios, et. al. "Pests, Weeds, and Diseases in Agricultural Crop and Animal Husbandry Production" 2021, p. 135,
2 Howard, Jenny, "Dead Zones Explained." https://www.nationalgeographic.com/environment/article/dead-zones, July 31, 2019
3 Jessen, Christian. (2013). "Effects of Simulated Eutrophication and Overfishing on Coral Reef Invertebrates, Algae and Microbes in the Red Sea."
4 Noble Research Institute, LLC, "Why Are Nitrogen Prices So High?", https://www.noble.org/news/publications/ag-news-and-views/2001/april/why-are-nitrogen-prices-so-high/ 5/5/2023
5 Boerner, Leigh Krietsch, "Industrial ammonia production emits more CO2 than any other chemical-making reaction. Chemists want to change that", June 15, 2019, C&EN Chemical and Engineering News, Volume 97, Issue 24
6 Fields S., "Global nitrogen: cycling out of control." Environ Health Perspect. 2004 Jul;112(10):A556-63. doi: 10.1289/ehp.112-a556. PMID: 15238298; PMCID: PMC1247398.
7 Chengkai Qu, et. al, "Polyurethane Foam-Based Passive Air Samplers in Monitoring Persistent Organic Pollutants: Theory and Application" Environmental Geochemistry, 2018, Pages 521-542, https://doi.org/10.1016/B978-0-444-63763-5.00021-5.
https://www.sciencedirect.com/science/article/pii/B9780444637635000215
8 "Sandstorms Hit 400 Million People as China's 45 Year Anti-Desertification Efforts Fails," China Observer, https://youtu.be/8c5QpvCfbws Apr. 15th, 2023
9 Peterle, Tony J., "DDT in Antarctic Snow" Nature, Vol. 224, Nov. 8th, 1969

10 Li, J. et. al, "Observation of organochlorine pesticides in the air of the Mt. Everest region", Ecotoxicology and Environmental Safety, Volume 63, Issue 1, January 2006, Pages 33-41, https://doi.org/10.1016/j.ecoenv.2005.04.001

11 Ivantsova, Emma, et. al, "Developmental and behavioral toxicity assessment of glyphosate and its main metabolite aminomethylphosphonic acid (AMPA) in zebrafish embryos/larvae" Environmental Toxicology and Pharmacology Volume 93, July 2022, 103873, https://doi.org/10.1016/j.etap.2022.103873

12 Gizeuda de F Sousa, et. al, "Evaluation of the atmospheric contamination level for the use of herbicide glyphosate in the northeast region of Brazil", Environ. Monit. Assess., 2019 Sep 4;191(10):604. doi: 10.1007/s10661-019-7764-x.

13 Alonso, Lucas L., et al. "Glyphosate and atrazine in rainfall and soils in agroproductive areas of the pampas region in Argentina" Science of The Total Environment, Volume 645, 15 December 2018, Pages 89-96

14 Chang, F.-c., Simcik, M.F. and Capel, P.D. (2011), Occurrence and fate of the herbicide glyphosate and its degradate aminomethylphosphonic acid in the atmosphere. Environmental Toxicology and Chemistry, 30: 548-555. https://doi.org/10.1002/etc.431

15 Gillam, Carey "U.S. researchers find Roundup chemical in water, air", Reuters, August 31, 2011 2:05 PM, 5/4/2023

16 Kontogiannatos, Dimitrios, et. al. "Pests, Weeds, and Diseases in Agricultural Crop and Animal Husbandry Production" 2021, p. 138

8. APEX OMNIVORE

1 Francis Chaboussou, "Healthy Crops: A New Agricultural Revolution" 1984, Jon Carpenter Publishing, p 51

2 Smith, Jeffery, M., "Genetic Roulette: The Documented Health Risks of Genetically Engineered Foods,", Yes! Books, 2007, p. 2

3 Ibid p. 129

4 Seneff, Stephanie & Swanson, Nancy & Li, Chen. (2015). "Aluminum and Glyphosate Can Synergisti-cally Induce Pineal Gland Pathology: Con-nection to Gut Dysbiosis and Neurological Disease." Agricultural Sciences. 06. 42-70. 10.4236/as.2015.61005.

5 Strilbyska OM, Tsiumpala SA, Kozachyshyn II, Strutynska T, Burdyliuk N, Lushchak VI, Lushchak O. "The effects of low-toxic herbicide Roundup and glyphosate on mitochondria." EXCLI J. 2022 Jan 10;21:183-196. doi: 10.17179/excli2021-4478. PMID: 35221840; PMCID: PMC8859649.

6 Ohn J. Fialka, "EPA Scientists Pressured to Allow Continued Use of Dangerous Pesticides," Wall Street Journal, May 25, 2006: A4, http://online.wsj.com/article/SB114852461658627 57.html

7 Skaper SD, Di Marzo V. "Endocannabinoids in nervous system health and disease: the big picture in a nutshell." Philos Trans R Soc Lond B Biol Sci. 2012 Dec 5;367(1607):3193-200. doi: 10.1098/rstb.2012.0313. PMID: 23108539; PMCID: PMC3481537.

8 Tallima, Hatem & El Ridi, Rashika. (2017). Arachidonic Acid: Physiological Roles and Potential Health Benefits. A Review.. Journal of Advanced Research. 11. 10.1016/j.jare.2017.11.004.

9 Yui K, Koshiba M, Nakamura S, Kobayashi Y. Effects of large doses of arachidonic acid added to docosahexaenoic acid on social impairment in individuals with autism spectrum disorders: a double-blind, placebo-controlled, randomized trial. J Clin Psychopharmacol. 2012 Apr;32(2):200-6. doi: 10.1097/JCP.0b013e3182485791. PMID: 22370992.

10 Zhang H, Cui Q, Song X. Research advances on arachidonic acid production by fermentation and genetic modification of Mortierella alpina. World J Microbiol Biotechnol. 2021 Jan 4;37(1):4. doi: 10.1007/s11274-020-02984-2. PMID: 33392832.

11 Naushad SM, Jain JM, Prasad CK, Naik U, Akella RR. "Autistic children exhibit distinct plasma amino acid profile." Indian J Biochem Biophys. 2013 Oct;50(5):474-8. PMID: 24772971.

12 Chen JR, Hsu SF, Hsu CD, Hwang LH, Yang SC. Dietary patterns and blood fatty acid composition in children with

attention-deficit hyperactivity disorder in Taiwan. J Nutr Biochem. 2004 Aug;15(8):467-72. doi: 10.1016/j.jnutbio.2004.01.008. PMID: 15302081.

13 Ito, Yuki, et. al, "Organophosphate Agent Induces ADHD-Like Behaviors via Inhibition of Brain Endocannabinoid-Hydrolyzing Enzyme(s) in Adolescent Male Rats" J. Agric. Food Chem. 2020, 68, 8, 2547–2553, Publication Date:January 30, 2020 https://doi.org/10.1021/acs.jafc.9b08195

14 Waits A, Chang CH, Yu CJ, Du JC, Chiou HC, Hou JW, Yang W, Chen HC, Chen YS, Hwang B, Chen ML. "Exposome of attention deficit hyperactivity disorder in Taiwanese children: exploring risks of endocrine-disrupting chemicals." J Expo Sci Environ Epidemiol. 2022 Jan;32(1):169-176. doi: 10.1038/s41370-021-00370-0. Epub 2021 Jul 15. PMID: 34267309.

15 Stern, Carly M., "A Slow-Motion Crisis: Gen Z's Battle Against Depression, Addiction, Hopelessness" September 7, 2022 https://www.the74million.org/article/a-slow-motion-crisis-gen-zs-battle-against-depression-addiction-hopelessness/ 5/6/2023

16 Bethune, Sophie, "Gen Z more likely to report mental health concerns" January 2019, Vol 50, No. 1, 5/6/2023, https://www.apa.org/monitor/2019/01/gen-z

17 Young SN. "Acute tryptophan depletion in humans: a review of theoretical, practical and ethical aspects." J Psychiatry Neurosci. 2013 Sep;38(5):294-305. doi: 10.1503/jpn.120209. PMID: 23428157; PMCID: PMC3756112.

18 Aitbali Y, Ba-M'hamed S, Elhidar N, Nafis A, Soraa N, Bennis M. "Glyphosate based- herbicide exposure affects gut microbiota, anxiety and depression-like behaviors in mice." Neurotoxicol Teratol. 2018 May-Jun;67:44-49. doi: 10.1016/j.ntt.2018.04.002. Epub 2018 Apr 7. PMID: 29635013.

19 Mineur, Yann S., "Cholinergic signaling in the hippocampus regulates social stress resilience and anxiety- and depression-like behavior" February 11, 2013 https://www.pnas.org/doi/10.1073/pnas.1219731110, 5/6/2023

20 Dulawa SC, Janowsky DS. "Cholinergic regulation of mood: from basic and clinical studies to emerging therapeutics." Mol Psychiatry. 2019 May;24(5):694-709. doi: 10.1038/s41380-018-0219-x. Epub 2018 Aug 17. PMID: 30120418; PMCID: PMC7192315.

21 Ait-Bali Y, Ba-M'hamed S, Gambarotta G, Sassoè-Pognetto M, Giustetto M, Bennis M. "Pre- and postnatal exposure to glyphosate-based herbicide causes behavioral and cognitive impairments in adult mice: evidence of cortical ad hippocampal dysfunction." Arch Toxicol. 2020 May;94(5):1703-1723. doi: 10.1007/s00204-020-02677-7. Epub 2020 Feb 17. PMID: 32067069.

22 Cattani D, Cesconetto PA, Tavares MK, Parisotto EB, De Oliveira PA, Rieg CEH, Leite MC, Prediger RDS, Wendt NC, Razzera G, Filho DW, Zamoner A. "Developmental exposure to glyphosate-based herbicide and depressive-like behavior in adult offspring: Implication of glutamate excitotoxicity and oxidative stress." Toxicology. 2017 Jul 15;387:67-80. doi: 10.1016/j.tox.2017.06.001. Epub 2017 Jun 13. PMID: 28627408.

23 "Stemming childhood obesity requires tackling root causes" Harvard T. H. Chan School of Public Health, https://www.hsph.harvard.edu/news/hsph-in-the-news/childhood-obesity-tackling-root-causes/, 5/6/2023

24 Remy Blumenfeld, "Gen Z At Work - 8 Reasons To Be Afraid" Oct 15, 2019, https://www.forbes.com/sites/remyblumenfeld/2019/10/15/gen-z-at-work8-reasons-to-be-afraid/?sh=4084d54133a9, 5/6/2023

25 Smith, Jeffery M., "Genetic Roulette: The Documented Health Risks of Genetically Engineered Foods," Yes! Books, 2007, p. 49

26 Giulia Carbonaro, "Number of Gen Z People Identifying as Transgender Twice That of Millennials" 02/24/23 https://www.newsweek.com/people-who-identify-transgender-doubles-gen-z-1783562 5/6/23

27 Megan Hull, "Gender Dysphoria Statistics" May 26, 2022, https://www.therecoveryvillage.com/mental-health/gender-dysphoria/gender-dysphoria-statistics/, 5/6/2023

28 Tom Costello, "Male fish becoming female?" Nov. 8, 2004, https://www.nbcnews.com/id/wbna6436617, 5/6/23

29 Mnif W, Hassine AI, Bouaziz A, Bartegi A, Thomas O, Roig B. "Effect of endocrine disruptor pesticides: a review." Int J Environ Res Public Health. 2011 Jun;8(6):2265-303. doi: 10.3390/ijerph8062265. Epub 2011 Jun 17. PMID: 21776230; PMCID: PMC3138025.

30 Smith, Jeffery M., "Genetic Roulette: The Documented Health Risks of Genetically Engineered Foods," Yes! Books, 2007, p. 37

31 Brenda Goodman, "Sperm counts may be declining globally, review finds, adding to debate over male fertility" November 18, 2022 https://www.cnn.com/2022/11/18/health/sperm-counts-decline-debate/index.html, 5/8/23

32 Dr Mae-Wan Ho, "Glyphosate/Roundup & Human Male Infertility", March 21, 2014, https://www.permaculturenews.org/2014/03/21/glyphosate-roundup-human-male-infertility/, 5/8/23

33 "Early-Onset Dementia and Alzheimer's Diagnoses Spiked 373 Percent for Generation X and Millennials", Blue Cross Blue Shield, Press Release, Feb. 27, 2020, https://www.bcbs.com/press-releases/early-onset-dementia-and-alzheimers-diagnoses-spiked-373-percent-generation-x-and-millennials 5/8/23

34 "Alzheimer's Disease Mortality by State", CDC/National Center for Health Statistics, February 28, 2022 https://www.cdc.gov/nchs/pressroom/sosmap/alzheimers_mortality/alzheimers_disease.htm, 5/9/23

35 Shi, L., Steenland, K., Li, H. et al. "A national cohort study (2000–2018) of long-term air pollution exposure and incident dementia in older adults in the United States." Nat Commun 12, 6754 (2021). https://doi.org/10.1038/s41467-021-27049-2

36 Gong, Y., Zhang, X., Zhao, X. et al. "Global ambient particulate matter pollution and neurodegenerative disorders: a systematic review of literature and meta-analysis." Environ Sci Pollut Res 30, 39418–39430 (2023). https://doi.org/10.1007/s11356-023-25731-0

9. FORESTS OF ARABIA

1 Meredith Brand, "Interview: Unraveling Arabia's green past" 9, April 2018, https://www.natureasia.com/en/nmiddleeast/article/10.1038/nmiddleeast.2018.44, 5/10/23
2 Carly Cassella, 12 December 2022, "Mysterious Patterns Span The Arabian Desert, And We May Finally Know Why," https://www.sciencealert.com/mysterious-patterns-span-the-arabian-desert-and-we-may-finally-know-why, 5/10/23
3 Sowell, Thomas, "Black Rednecks and White Liberals," 2005, p. 130
4 "Buying Time," 2013 Smithsonian, https://smithsonianeducation.org/educators/lesson_plans/revolutionary_money/intro_3.html, 5/20/23
5 Worster, Donald, "Dust Bowl," 1952, Texas State Historical Association, https://www.tshaonline.org/handbook/entries/dust-bowl, 5/18/23
6 Lotterman, Edward, "The Porcine Slaughter of the Innocents," 1/1/1999, https://www.minneapolisfed.org/article/1999/the-porcine-slaughter-of-the-innocents, 5/20/23
7 Justice Roberts, et. al, "United States v. Butler," 1/6/1936, https://www.law.cornell.edu/supremecourt/text/297/1, 5/20/23
8 Edwards, Chris, "Agricultural Subsidies," April 16, 2018, https://www.downsizinggovernment.org/agriculture/subsidies
9 Drisket, Tiarra, "Farm Subsidies: Harmful or Helpful?" 11/3/21, Texas A&M, College of Liberal Arts, https://liberalarts.tamu.edu/blog/2021/11/03/farm-subsidies-harmful-or-helpful/, 5/22/23

10. ALL THINGS NEWS

1 Waldeck, Sabine, "US suffers a severe decline in organic farming transition since 2008, new initiative poised to reverse trend"

Food Ingredients First , 08/23/22, https://www.foodingredientsfirst.com/news/us-suffers-a-severe-decline-in-organic-farming-transition-since-2008-new-initiative-poised-to-reverse-trend.html, 5/25/23

2 USDA, ERS, "The Role of Changing Vertical Coordination in the Broiler and Pork Industries" https://www.ers.usda.gov/webdocs/publications/40999/17957_aer777b_1_.pdf?v=0, 5/25/23

BIBLIOGRAPHY

A.L. Elliott, J.G. Davis, R.M. Waskom, J.R. Self and D.K. Christensen,* *"Phosphorus Fertilizers for Organic Farming Systems – 0.569"* Colorado State University Extension, 2014 https://extension.colostate.edu/topic-areas/agriculture/phosphorus-fertilizers-for-organic-farming-systems-0-569/

Abdel-Azeem, Ahmed & Mansour, Samira. (2015). *A new record of Taxonomy and ecology of antibiotic producing Actinomycetes.* African journal of agricultural research. 7. 2255-2261. 10.5897/AJARX11.071.

Adetunji, Jacob, *Explainer: what dust from the Sahara does to you and the planet,* The Conversation, April 20, 2016, https://theconversation.com/explainer-what-dust-from-the-sahara-does-to-you-and-the-planet-57373

Advancing Earth and Space Science, *Aerosol pollution caused decades of 'global dimming',* 2021, https://news.agu.org/press-release/aerosol-pollution-caused-decades-of-global-dimming/

Ahl, Jonathan, *EPA reverses course on herbicide ban in Missouri and 7 other states,* NPR, 2022, https://news.stlpublicradio.org/health-science-environment/2022-04-04/epa-reverses-course-on-herbicide-ban-in-missouri-and-7-other-states

Ait-Bali Y, Ba-M'hamed S, Gambarotta G, Sassoè-Pognetto M, Giustetto M, Bennis M. Pre- and postnatal exposure to glyphosate-based herbicide causes behavioral and cognitive impairments in adult mice: evidence of cortical ad hippocampal dysfunction. Arch Toxicol. 2020 May;94(5):1703-1723. doi: 10.1007/s00204-020-02677-7. Epub 2020 Feb 17. PMID: 32067069.

Albeck-Ripka, Livia, "Faced With Drought, the Pharaohs Tried (and Failed) to Adapt" The New York Times, March 30, 2018 https://www.nytimes.com/2018/03/30/climate/egypt-climate-drought.html

Al-Kaisi, Mahdi, "Fall Versus Spring Tillage, Which is Better", Iowa State University, Extension and Outreach, September 23, 2010, https://crops.extension.iastate.edu/cropnews/2010/09/fall-versus-spring-tillage-which-better

Allen, J, "Galls & Burls Explained...Sort of", University of Connecticut Home and Garden Education Center, April 20, 2011, https://uconnladybug.wordpress.com/2011/04/20/galls-burls-explained-sort-of/

Aloui A, Recorbet G, Gollotte A, Robert F, Valot B, Gianinazzi-Pearson V, Aschi-Smiti S, Dumas-Gaudot E. "On the mechanisms of cadmium stress alleviation in Medicago truncatula by arbuscular mycorrhizal symbiosis: a root proteomic study." Proteomics. 2009 Jan;9(2):420-33. doi: 10.1002/pmic.200800336. PMID: 19072729.

Alqahtani S, Butcher MC, Ramage G, Dalby MJ, McLean W, Nile CJ. "Acetylcholine Receptors in Mesenchymal Stem Cells." Stem Cells Dev. 2023 Feb;32(3-4):47-59. doi: 10.1089/scd.2022.0201. Epub 2023 Jan 24. PMID: 36355611.

American Phytopathological Society, "Scientists compare soil microbes in no-till, conventional tilling systems of Pacific Northwest farms" August 17, 2017, https://phys.org/news/2017-08-scientists-soil-microbes-no-till-conventional.html

America's Wetland Foundation Resource Center. "Mississippi River Anatomy" http://www.americaswetlandresources.com/background_facts/detailedstory/MississippiRiverAnatomy.html.

Ancient Ports - Ports Antiques, "Leptis Manga", https://www.ancientportsantiques.com/a-few-ports/leptis-magna/, 2017

Antezana PE, Colin VL, Bourguignon N, Benimeli CS, Fuentes MS. "Applied of actinobacteria consortia-based bioremediation to restore co-contaminated systems." Res Microbiol. 2023 May;174(4):104028. doi: 10.1016/j.resmic.2023.104028. Epub 2023 Jan 11. PMID: 36638934.

Armitage, Simon J., Bristow, Charlie S., and Drake, Nick A., "West African monsoon dynamics inferred from abrupt fluctuations of Lake Mega-Chad", Proceedings of the National Academy of Sciences (PNAS) June 29, 2015, https://doi.org/10.1073/pnas.1417655112

Arnason, Robert, "Survey Reveals 'Amazing' Soil Loss in Great Plains Region" The Western Producer, Jan. 19th, 2017, https://www.producer.com/2017/01/survey-reveals-amazing-soil-loss-in-great-plains-region

Aulakh, R., Gill, J., Bedi, J., Sharma, J., Joia, B., & Ockerman, H.. "Organochlorine pesticide residues in poultry feed, chicken muscle and eggs at a poultry farm in Punjab, India." Journal of the science of food and agriculture, 86, 741-744. doi: 10.1002/jsfa.2407

Axelsson, S. (1991), "Origin and Significance of Acetylcholine and Choline in Plasma and Serum from Normal and Paretic Cows." Journal of Veterinary Medicine Series A, 38: 737-748. https://doi.org/10.1111/j.1439-0442.1991.tb01073.x

Azziz, R., Carmina, E., Chen, Z. et al. "Polycystic ovary syndrome." Nat Rev Dis Primers 2, 16057 (2016). https://doi.org/10.1038/nrdp.2016.57

Badarch, D., Zilinskas, R.A., "Mongolia Today Science, Culture, Environment and Development", Routledge 2015

Baiocchi, T., Lee, G., Choe, DH. et al. "Host seeking parasitic nematodes use specific odors to assess host resources." Sci Rep

7, 6270 (2017). https://doi.org/10.1038/s41598-017-06620-2

Baldi, Elisabetta, Keshani, Parisa, Sharifi, Mohammad Hossein, Heydari, Mohammad Reza, Joulaei, Hassan, "The Effect of Genetically Modified Food on Infertility Indices: A Systematic Review Study", The Scientific World Journal, 2020, https://doi.org/10.1155/2020/1424789

Barman, J., Samanta, A., Saha, B. et al. "Mycorrhiza." Reson 21, 1093–1104 (2016). https://doi.org/10.1007/s12045-016-0421-6

Barnett JA, Bandy ML, Gibson DL. "Is the Use of Glyphosate in Modern Agriculture Resulting in Increased Neuropsychiatric Conditions Through Modulation of the Gut-brain-microbiome Axis?" Frontiers in Nutrition. 2022 ;9:827384. DOI: 10.3389/fnut.2022.827384. PMID: 35356729; PMCID: PMC8959108.

Bastos Sales L, Kamstra JH, Cenijn PH, van Rijt LS, Hamers T, Legler J. "Effects of endocrine disrupting chemicals on in vitro global DNA methylation and adipocyte differentiation." Toxicol In Vitro. 2013 Sep;27(6):1634-43. doi: 10.1016/j.tiv.2013.04.005. Epub 2013 Apr 18. PMID: 23603478.

Baurick, Tristan, "Gulf dead zone inflicts billions in damage to Louisiana fisheries and environment, report says" Times-Picayune | New Orleans Advocate, 2020, https://www.houmatoday.com/story/news/2020/06/09/gulf-dead-zone-inflicts-billions-in-damage/112590036/

Beard JD, Steege AL, Ju J, Lu J, Luckhaupt SE, Schubauer-Berigan MK. "Mortality from Amyotrophic Lateral Sclerosis and Parkinson's Disease Among Different Occupation Groups - United States, 1985-2011." MMWR Morb Mortal Wkly Rep. 2017 Jul 14;66(27):718-722. doi: 10.15585/mmwr.mm6627a2. PMID: 28704346; PMCID: PMC5687590.

Benbrook CM. Trends in glyphosate herbicide use in the United States and globally. Environ Sci Eur. 2016;28(1):3. doi: 10.1186/s12302-016-0070-0. Epub 2016 Feb 2. PMID: 27752438; PMCID: PMC5044953.

Benbrook, C.M. "Impacts of genetically engineered crops on pesticide use in the U.S. -- the first sixteen years." Environ Sci Eur 24, 24 (2012). https://doi.org/10.1186/2190-4715-24-24

Bentley, S.D. & Chater, K.F. & Cerdeño-Tárraga, A.-M & Challis, G.L. & Thomson, N.R. & James, K.D. & Harris, D.E. & Quail, M.A. & Kieser, H & Harper, Delphine & Bateman, Alex & Brown, S & Chandra, Govind & Chen, Carton & Collins, M & Cronin, Ann & Fraser, A & Goble, A & Hidalgo, Jhorhi & Hopwood, David. (2002). Complete genome sequence of the model actinomycete Streptomyces coelicolor A3(2). Nature. 417. 141-7. 10.1038/417141a.

Ben-Yehoshua, Shimshon & Borowitz, Carole & Hanus, Lumir. (2012). "Frankincense, Myrrh, and Balm of Gilead: Ancient Spices of Southern Arabia and Judea." Horticultural Reviews. 39. 1-76. 10.1002/9781118100592.ch1.

Berardelli, Jeff, "A devastating Dust Bowl heat wave is now more than twice as likely, study says" CBS News, 2020, https://www.cbsnews.com/news/dust-bowl-heat-wave-climate-change-twice-as-likely-study-says/

Bethune, Sophie, "Gen Z more likely to report mental health concerns", American Psychological Association, 2019, https://www.apa.org/monitor/2019/01/gen-z

Bettelherim, Frederick A., and March, Jerry, "Introduction to Organic & Biochimistery" Harcourt Brace College Publishers, 1990

Beven, Keith & Brazier, R.. (2011). "Dealing with Uncertainty in Erosion Model Predictions" https://doi.org10.1002/9781444328455.ch4.

Beyond Pesticides, "From Udder to Table: Toxic Pesticides Found in Conventional Milk, Not Organic Milk", July, 9th, 2020, https://beyondpesticides.org/dailynewsblog/2020/07/from-udder-to-table-toxic-pesticides-found-in-conventional-milk-not-organic-milk/

Biehl, Michael L., Buck, William B., "Chemical Contaminants: Their Metabolism and their Residues," Journal of Food Protection, Vol. 50, 1987, Pgs. 1058-1073, https://doi.org/10.4315/0362-028X-50.12.1058.

Blaikie, Piers, "The Political Economy of Soil Erosion in Developing Countries" Routledge, 1985

Block, Jennifer, "Gender dysphoria in young people is rising—and so is professional disagreement", BMJ 2023; 380 doi: https://doi.org/10.1136/bmj.p382

Blot N, Veillat L, Rouzé R, Delatte H. "Glyphosate, but not its metabolite AMPA, alters the honeybee gut microbiota." PLoS One. 2019 Apr 16;14(4):e0215466. doi: 10.1371/journal.pone.0215466. PMID: 30990837; PMCID: PMC6467416.

Blumenfeld, Remy, "Gen Z At Work - 8 Reasons To Be Afraid" Forbes, 2019, https://www.forbes.com/sites/remyblumenfeld/2019/10/15/gen-z-at-work8-reasons-to-be-afraid/?sh=4084d54133a9

Boerboom, Chris, and Owen, Michael, "Facts About Glyphosate Resistant Weeds", Purdue University, 2006, https://www.extension.purdue.edu/extmedia/gwc/gwc-1.pdf

Bogaert, Lies, Martens, Ann, Wijbe Kast, Martin, Van Marck, Eric, De Cock, Hilde, "Bovine papillomavirus DNA can be detected in keratinocytes of equine sarcoid tumors", Veterinary Microbiology, Vol. 146, 2010, Pgs. 269-275, https://doi.org/10.1016/j.vetmic.2010.05.032.

Borrelli, P., Lugato, E., Montanarella, L., and Panagos, P. (2017) "A New Assessment of Soil Loss Due to Wind Erosion in European Agricultural Soils Using a Quantitative Spatially Distributed Modelling Approach." Land Degrad. Develop., 28: 335– 344. doi: 10.1002/ldr.2588.

Brand, Meredith, "Interview: Unraveling Arabia's Green Past", Nature Middle East, 2018, https://www.natureasia.com/en/nmiddleeast/article/10.1038/nmiddleeast.2018.44

Brausch, John & Smith, Philip. (2009). "Mechanisms of resistance and cross-resistance to agrochemicals in the fairy shrimp Thamnocephalus platyurus (Crustacea: Anostraca)." Aquatic toxicology (Amsterdam, Netherlands). 92. 140-5. 10.1016/j.aquatox.2009.02.002.

Bretveld RW, Thomas CM, Scheepers PT, Zielhuis GA, Roeleveld N. "Pesticide exposure: the hormonal function of the female reproductive system disrupted?" Reproductive Biology and

Endocrinology : RB&E. 2006 May;4:30. DOI: 10.1186/1477-7827-4-30. PMID: 16737536; PMCID: PMC1524969.

Browning M, Dawson C, Alm SR, McElderry CF, Amador JA. "Effect of Carbon Amendment and Soil Moisture on Tylenchorhynchus spp. and Hoplolaimus galeatus." J Nematol. 1999 Dec;31(4):445-54. PMID: 19270917; PMCID: PMC2620395.

Bruinenberg, M.H. & Agtmaal, Maaike & Hoekstra, Nyncke & van Eekeren, Nick. (2023). "Residues of pesticides in dairy cow rations and fly treatments reduce the number of Coleoptera in dung." Agriculture, Ecosystems & Environment. 344. 108307. 10.1016/j.agee.2022.108307.

Buehring GC, Sans HM. "Breast Cancer Gone Viral? Review of Possible Role of Bovine Leukemia Virus in Breast Cancer, and Related Opportunities for Cancer Prevention." Int J Environ Res Public Health. 2019 Dec 27;17(1):209. doi: 10.3390/ijerph17010209. PMID: 31892207; PMCID: PMC6982050.

Bueno de Mesquita, Clifton P & Solon, Adam & Barfield, Amy & Mastrangelo, Claire & Tubman, Abigail & Vincent, Kim & Porazinska, Dorota & Hufft, Rebecca & Shackelford, Nancy & Suding, Katharine & Schmidt, Steven. (2023). "Adverse impacts of Roundup on soil bacteria, soil chemistry and mycorrhizal fungi during restoration of a Colorado grassland." Applied Soil Ecology. 185. 104778. 10.1016/j.apsoil.2022.104778.

Buman, Tom, "Cost of soil erosion in Iowa is not a pretty picture", https://www.agrentools.com

Burkholder J, Libra B, Weyer P, Heathcote S, Kolpin D, Thorne PS, Wichman M. "Impacts of waste from concentrated animal feeding operations on water quality. Environ Health Perspect." 2007 Feb;115(2):308-12. doi: 10.1289/ehp.8839. Epub 2006 Nov 14. PMID: 17384784; PMCID: PMC1817674.

Business Day, "UNESCO supports FG to save Lake Chad from extinction", Nov 10, 2017, https://businessday.ng/uncategorized/article/unesco-supports-fg-save-lake-chad-extinction/

Cai S, Kumar R, Singh BR. "Clostridial Neurotoxins: Structure, Function and Implications to Other Bacterial Toxins." Microorganisms. 2021 Oct 23;9(11):2206. doi:

10.3390/microorganisms9112206. PMID: 34835332; PMCID: PMC8618262.

Canfield, Donald & Glazer, Alexander & Falkowski, Paul. (2010). "The Evolution and Future of Earth's Nitrogen Cycle." Science (New York, N.Y.). 330. 192-6. 10.1126/science.1186120.

Cappellozza BI, Cooke RF, Harvey KM. "Omega-6 Fatty Acids: A Sustainable Alternative to Improve Beef Production Efficiency." Animals (Basel). 2021 Jun 12;11(6):1764. doi: 10.3390/ani11061764. PMID: 34204706; PMCID: PMC8231484.

Carrageta, DF, Oliveira, PF, Alves, MG, Monteiro, MP. "Obesity and male hypogonadism: Tales of a vicious cycle." Obesity Reviews. 2019; 20: 1148– 1158. https://doi.org/10.1111/obr.12863

Casida, John E. "Chapter 2 - Pest Toxicology: The Primary Mechanisms of Pesticide Action," Hayes' Handbook of Pesticide Toxicology (Third Edition), Academic Press, 2010, Pgs. 103-117, https://doi.org/10.1016/B978-0-12-374367-1.00002-1.

Cassella, Carly, "Mysterious Patterns Span The Arabian Desert, And We May Finally Know Why", Science Alert, 2022, https://www.sciencealert.com/mysterious-patterns-span-the-arabian-desert-and-we-may-finally-know-why

Cattani D, Cesconetto PA, Tavares MK, Parisotto EB, De Oliveira PA, Rieg CEH, Leite MC, Prediger RDS, Wendt NC, Razzera G, Filho DW, Zamoner A. "Developmental exposure to glyphosate-based herbicide and depressive-like behavior in adult offspring: Implication of glutamate excitotoxicity and oxidative stress." Toxicology. 2017 Jul 15;387:67-80. doi: 10.1016/j.tox.2017.06.001. Epub 2017 Jun 13. PMID: 28627408.

Cattani D, Struyf N, Steffensen V, Bergquist J, Zamoner A, Brittebo E, Andersson M. "Perinatal exposure to a glyphosate-based herbicide causes dysregulation of dynorphins and an increase of neural precursor cells in the brain of adult male rats." Toxicology. 2021 Sep;461:152922. doi: 10.1016/j.tox.2021.152922. Epub 2021 Aug 30. PMID: 34474092.

Center for Biological Diversity, "Endocrine-disrupting Pesticide Atrazine to Be Banned in Hawaii, Five U.S. Territories, Prohibited on Conifers, Roadsides", 2020, https://biologicaldiversity.org/w/news/press-releases/endocrine-disrupting-pesticide-atrazine-be-banned-hawaii-five-us-territories-prohibited-conifers-roadsides-2020-09-23/

Cervera, Marc, "Organically off-target: EU plans fall short for countries to achieve 25% organic farms by 2030", Food Ingredients First, https://www.foodingredientsfirst.com/news/organically-off-target-eu-plans-fall-short-for-countries-to-achieve-25-organic-farms-by-2030.html

Chaboussou, Francis, "Heathly Crops", Jon Carpenter Publishing, 2004

Chang FC, Simcik MF, Capel PD. "Occurrence and fate of the herbicide glyphosate and its degradate aminomethylphosphonic acid in the atmosphere." Environ Toxicol Chem. 2011 Mar;30(3):548-55. doi: 10.1002/etc.431. Epub 2011 Jan 19. PMID: 21128261.

Charles, Dan, "As Biotech Crops Lose Their Power, Scientists Push For New Restrictions", NPR, October 29, 2020, https://www.npr.org/2020/10/29/927111009/as-biotech-crops-lose-their-power-scientists-push-for-new-restrictions

Chen M, Arato M, Borghi L, Nouri E, Reinhardt D. "Beneficial Services of Arbuscular Mycorrhizal Fungi - From Ecology to Application" Front Plant Sci. 2018 Sep 4;9:1270. doi: 10.3389/fpls.2018.01270. PMID: 30233616; PMCID: PMC6132195.

Chen, Baodong. (2006). Humic Acids Increase the Phytoavailability of Cd and Pb to Wheat Plants Cultivated in Freshly Spiked, Contaminated Soil (7 pp). Journal of Soils and Sediments. 6. 236-242. 10.1065/jss2006.08.178.

Chin JY, Godwin C, Jia C, Robins T, Lewis T, Parker E, Max P, Batterman S. "Concentrations and risks of p-dichlorobenzene in indoor and outdoor air." Indoor Air. 2013 Feb;23(1):40-9. doi: 10.1111/j.1600-0668.2012.00796.x. Epub 2012 Jul 18. PMID: 22725685; PMCID: PMC3501547.

Chuntao Yin et al, "Bacterial Communities on Wheat Grown Under Long-Term Conventional Tillage and No-Till in the Pacific Northwest of the United States," Phytobiomes (2017). DOI: 10.1094/PBIOMES-09-16-0008-R

Clark DE, Palmer JS, Radeleff RD, Crookshank HR, Farr FM. "Residues of chlorophenoxy acid herbicides and their phenolic metabolites in tissues of sheep and cattle." J Agric Food Chem. 1975 May-Jun;23(3):573-8. doi: 10.1021/jf60199a032. PMID: 1151004.

Cleveland Clinic, "Dopamine", 2023, https://my.clevelandclinic.org/health/articles/22581-dopamine

Colović MB, Krstić DZ, Lazarević-Pašti TD, Bondžić AM, Vasić VM. "Acetylcholinesterase inhibitors: pharmacology and toxicology." Curr Neuropharmacol. 2013 May;11(3):315-35. doi: 10.2174/1570159X11311030006. PMID: 24179466; PMCID: PMC3648782.

Compart, Devan P., "Targeting Amino Acids in Beef Cattle Diets Could Rumen-Protected Amino Acids Improve Performance and Reduce Waste?" Feedlot, Aug 21, 2021, https://www.feedlotmagazine.com/news/feedlot_special/targeting-amino-acids-in-beef-cattle-diets-could-rumen-protected-amino-acids-improve-performance-and/article_7dada186-eb1f-11eb-8e75-0b50cc241cd6.html

Compston JE, Vedi S, Stephen AB, Bord S, Lyons AR, Hodges SJ, Scammell BE. "Reduced bone formation after exposure to organophosphates." Lancet. 1999 Nov 20;354(9192):1791-2. doi: 10.1016/s0140-6736(99)04466-9. PMID: 10577647.

Compton, Jana & Boone, Richard. (2000). "Long-Term Impacts of Agriculture on Soil Carbon and Nitrogen in New England Forests." Ecology. 81. 2314-2330. 10.1890/0012-9658(2000)081[2314:LTIOAO]2.0.CO;2.

Congressional Research Service, "Farm Bill Primer: PLC and ARC Farm Support Programs", 2022, https://crsreports.congress.gov/product/pdf/IF/IF12114

Coninx, Laura, Martinova, Veronika, Rineau, Francois, "Chapter Four - Mycorrhiza-Assisted Phytoremediation" Advances in Botanical Research, Phytoremediation, Pages 127-188 (2017)

Connolly A, Leahy M, Jones K, Kenny L, Coggins MA. "Glyphosate in Irish adults - A pilot study in 2017." Environ Res. 2018 Aug;165:235-236. doi: 10.1016/j.envres.2018.04.025. Epub 2018 May 3. PMID: 29729481.

Coperchini F, Greco A, Croce L, Denegri M, Magri F, Rotondi M, Chiovato L. "In vitro study of glyphosate effects on thyroid cells." Environ Pollut. 2023 Jan 15;317:120801. doi: 10.1016/j.envpol.2022.120801. Epub 2022 Nov 30. PMID: 36462676.

Corn and Soybean Digest "Soil Wealth, Why North America Feeds World", 2017, https://www.cornandsoybeandigest.com/issues/soil-wealth-why-north-america-feeds-world

Cosme, Nuno & Hauschild, Michael. (2017). "Characterization of waterborne nitrogen emissions for marine eutrophication modelling in life cycle impact assessment at the damage level and global scale." The International Journal of Life Cycle Assessment. 22. 10.1007/s11367-017-1271-5.

Costas-Ferreira C, Durán R, Faro LRF. "Toxic Effects of Glyphosate on the Nervous System: A Systematic Review." Int J Mol Sci. 2022 Apr 21;23(9):4605. doi: 10.3390/ijms23094605. PMID: 35562999; PMCID: PMC9101768.

Costello, Tom, "Male fish becoming female?", NBC, 2004, https://www.nbcnews.com/id/wbna6436617

Cot, A. L. (2005). "Breed out the Unfit and Breed in the Fit": Irving Fisher, Economics, and the Science of Heredity. The American Journal of Economics and Sociology, 64(3), 793–826. http://www.jstor.org/stable/3488160

Couvidat, Florian & Bedos, Carole & Gagnaire, Nathalie & Carra, Mathilde & Ruelle, Bernadette & Martin, Philippe & Poméon, Thomas & Alletto, Lionel & Armengaud, Alexandre & Quivet, Etienne. (2021). "Simulating the impact of volatilization on atmospheric concentrations of pesticides with the 3D chemistry-transport model CHIMERE: Method development and application to S-metolachlor and folpet." Journal of Hazardous Materials. 424. 10.1016/j.jhazmat.2021.127497.

Cox C, Surgan M. "Unidentified inert ingredients in pesticides: implications for human and environmental health." Environ

Health Perspect. 2006 Dec;114(12):1803-6. doi: 10.1289/ehp.9374. PMID: 17185266; PMCID: PMC1764160.

Cox GB, Gibson F. "The role of shikimic acid in the biosynthesis of vitamin K2." Biochem J. 1966 Jul;100(1):1-6. doi: 10.1042/bj1000001. PMID: 5337721; PMCID: PMC1265084.

Craine, J.M., Brookshire, E.N.J., Cramer, M.D. et al. "Ecological interpretations of nitrogen isotope ratios of terrestrial plants and soils." Plant Soil 396, 1–26 (2015). https://doi.org/10.1007/s11104-015-2542-1

Creech, Elizabeth, "Saving Money, Time, and Soil: The Economics of No-Till Farming" NRCS, Nov. 30th, 2017, https://www.usda.gov/media/blog/2017/11/30/saving-money-time-and-soil-economics-of-no-till-farming

Currie, C., Scott, J., Summerbell, R. et al. Fungus-growing ants use antibiotic-producing bacteria to control garden parasites. Nature 398, 701–704 (1999). https://doi.org/10.1038/19519

Davis, Shemina, "Soil loss—an unfolding global disaster" Phys.org., December 2, 2015, https://phys.org/news/2015-12-soil-lossan-unfolding-global-disaster.html

de Castro Vieira Carneiro, C. L., Chaves, E. M. C., Neves, K. R. T., Braga, M. D. M., Assreuy, A. M. S., de Moraes, M. E. A., & Aragão, G. F. (2023). "Behavioral and neuroinflammatory changes caused by glyphosate: Base herbicide in mice offspring. Birth Defects Research," 115(4), 488– 497. https://doi.org/10.1002/bdr2.2138

de Oliveira EP, Marchi KE, Emiliano J, Salazar SMCH, Ferri AH, Etto RM, Reche PM, Pileggi SAV, Kalks KHM, Tótola MR, Schemczssen-Graeff Z, Pileggi M. "Changes in fatty acid composition as a response to glyphosate toxicity in Pseudomonas fluorescens." Heliyon. 2022 Jul 13;8(8):e09938. doi: 10.1016/j.heliyon.2022.e09938. PMID: 35965982; PMCID: PMC9364109.

De Pol M, Kolla NJ. "Endocannabinoid markers in autism spectrum disorder: A scoping review of human studies." Psychiatry Res. 2021 Dec;306:114256. doi:

10.1016/j.psychres.2021.114256. Epub 2021 Oct 31. PMID: 34775294.

Defarge, Nicolas & Vendômois, Joël & Séralini, G.E.. (2017). "Toxicity of formulants and heavy metals in glyphosate-based herbicides and other pesticides." Toxicology Reports. 5. 10.1016/j.toxrep.2017.12.025.

DeFelice, Michael, "PPO Inhibitor (Cell Membrane Disruptor) Herbicides", Pioneer Crop Insights, 2023, https://www.pioneer.com/us/agronomy/ppo-inhibitor-herbicides.html

Deken, Dave, and Franklin, Jay, "No-Till Farming", SunupTV, 2010, https://www.youtube.com/watch?v=OqU0k_m6qAo

Del Castilo I, Neumann AS, Lemos FS, De Bastiani MA, Oliveira FL, Zimmer ER, Rêgo AM, Hardoim CCP, Antunes LCM, Lara FA, Figueiredo CP, Clarke JR. "Lifelong Exposure to a Low-Dose of the Glyphosate-Based Herbicide RoundUp® Causes Intestinal Damage, Gut Dysbiosis, and Behavioral Changes in Mice." Int J Mol Sci. 2022 May 17;23(10):5583. doi: 10.3390/ijms23105583. PMID: 35628394; PMCID: PMC9146949.

Devens, Eve and Hayes, Jared, "Nearly 20,000 farmers received farm subsidies for 37 consecutive years", Environmental Working Group, 2023, https://www.ewg.org/news-insights/news/2023/02/nearly-20000-farmers-received-farm-subsidies-37-consecutive-years

Devine, Jon, Baron, Valerie, Miller, D., Muren, Gregory, "CAFOs: What We Don't Know Is Hurting Us", NRDC, September 23, 2019, https://www.nrdc.org/resources/cafos-what-we-dont-know-hurting-us

Dhaliwal, G.S. & Jindal, Vikas & Dhawan, Ashok. (2010). Insect pest problems and crop losses: Changing trends. Indian J Ecol. 37. 1-7.

Dias MA, Batista PR, Ducati LC, Montagner CC. "Insights into sorption and molecular transport of atrazine, testosterone, and progesterone onto polyamide microplastics in different aquatic matrices." Chemosphere. 2023 Mar;318:137949. doi: 10.1016/j.chemosphere.2023.137949. Epub 2023 Jan 26. PMID: 36709842.

Dick, Richard P., "A review: long-term effects of agricultural systems on soil biochemical and microbial parameters," Agriculture, Ecosystems & Environment, Vol. 40, Issues 1–4, 1992, Pages 25-36, https://doi.org/10.1016/0167-8809(92)90081-L.

DiDonato, Nicole & Hatcher, Patrick. (2017). "Alicyclic carboxylic acids in soil humic acid as detected with ultrahigh resolution mass spectrometry and multi-dimensional NMR." Organic Geochemistry. 112. 10.1016/j.orggeochem.2017.06.010.

Dieter Jarasch, Dr. Ernst, "New pathogens in beef and cow's milk contributing to the risk of cancer", BIOPRO Baden-Württemberg GmbH, https://www.gesundheitsindustrie-bw.de/en/article/news/new-pathogens-in-beef-and-cows-milk-contributing-to-the-risk-of-cancer

Dirks, Harlan J., and Fienup, Darrell F., "Technological and market forces affecting Vertical Integration in the Hog Industry", University of Minnesota, Agricultural Experiment Station, Technical Bulletin 249, 1965, https://conservancy.umn.edu/bitstream/handle/11299/140012/1/TB249.pdf

Disabled World, "List of Generation Names by Year and Definition", 2023, https://www.disabled-world.com/definitions/generation-names.php

Donarummo, J., Ram, M., and Stoermer, E. F. (2003), "Possible deposit of soil dust from the 1930's U.S. dust bowl identified in Greenland ice," Geophys. Res. Lett., 30, 1269, doi:10.1029/2002GL016641, 6.

Donley, Nathan, "EPA Finds Atrazine Likely Harming Most Species of Plants, Animals in U.S." Center for Biological Diversity, May 3, 2016 https://www.biologicaldiversity.org/news/press_releases/2016/atrazine-05-03-2016.html

Doran, J.W. (1980), "Soil Microbial and Biochemical Changes Associated with Reduced Tillage." Soil Science Society of America Journal, 44: 765-771. https://doi.org/10.2136/sssaj1980.03615995004400040022x

Dorsey, E. Ray, Sherer, Todd, Okun, Michael S., Bloem, Bastiaan R., "The Rise of Parkinson's Disease", 2023, https://www.americanscientist.org/article/the-rise-of-parkinsons-disease

Dotson, Douglas, "Memorandum: PP#s 7F4841 and 0F6171 Tolerance Petitions for the Use of Flumioxazin on Peanuts, Soybeans, and Sugarcane, Evaluation of Residue Chemistry and Analytical Methodology", EPA: Office of Prevention, Pesticides, and Toxic Substances, 3/12/2001, https://www3.epa.gov/pesticides/chem_search/cleared_reviews/csr_PC-129034_12-Mar-01_a.pdf

Drijber, Rhae & Doran, John & Parkhurst, Anne & Lyon, Drew. (2000). "Changes in Soil Microbial Community Structure with Tillage Under Long-Term Wheat-Fallow Management." Soil Biology and Biochemistry. 32. 1419-1430. 10.1016/S0038-0717(00)00060-2.

Drinkwater, Laurie & Snapp, Sieglinde. (2007). "Understanding and Managing the Rhizosphere in Agroecosystems." https://www.doi.10.1016/B978-012088775-0/50008-2.

Drisker, Tiarra, "Farm Subsidies: Harmful or Helpful?", Texas A&M University, College of Arts and Sciences, 2021, https://liberalarts.tamu.edu/blog/2021/11/03/farm-subsidies-harmful-or-helpful/

Dulawa SC, Janowsky DS. "Cholinergic regulation of mood: from basic and clinical studies to emerging therapeutics." Mol Psychiatry. 2019 May;24(5):694-709. doi: 10.1038/s41380-018-0219-x. Epub 2018 Aug 17. PMID: 30120418; PMCID: PMC7192315.

Easy Chem, "History of the Haber Process" https://easychem.com.au/monitoring-and-management/maximising-production/history-of-the-haber-process/, 2017

Ebner, Paul, "CAFOs and Public Health", Purdue University, Purdue Extension, 8/2007,

Edwards, Chris, "Agricultural Subsidies", Downsizing The Federal Government, CATO Institute, 2018, https://www.downsizinggovernment.org/agriculture/subsidies

El-Tarabily KA, Nassar AH, Hardy GE, Sivasithamparam K. "Plant growth promotion and biological control of Pythium aphanidermatum, a pathogen of cucumber, by endophytic actinomycetes." J Appl Microbiol. 2009 Jan;106(1):13-26. doi: 10.1111/j.1365-2672.2008.03926.x. Epub 2008 Dec 12. Erratum in: J Appl Microbiol. 2009 Nov;107(5):1765-6. PMID: 19120624.

Encyclopedia of Life "Glomus mosseae" 2017 https://eol.org/pages988675/overview, https://eol.org/pages/307644

Encyclopedia.com, "Fisher, Irving", 2018, https://www.encyclopedia.com/people/social-sciences-and-law/economics-biographies/irving-fisher

Environmental Working Group, "Farm Subsidy Primer",2023, https://farm.ewg.org/subsidyprimer.php

EPA Pesticide Fact Sheet, "Name of Chemical: Carfentrazone-ethyl Reason for Issuance: New Chemical Registration Date Issued: Sept. 30th, 1998, https://www.epa.gov/pesticides/chem_search/reg_actions/registrations/fs_PC-128712_30-Sep-98.pdf

Erban T, Stehlik M, Sopko B, Markovic M, Seifrtova M, Halesova T, Kovaricek P. "The different behaviors of glyphosate and AMPA in compost-amended soil." Chemosphere. 2018 Sep;207:78-83. doi: 10.1016/j.chemosphere.2018.05.004. Epub 2018 May 4. PMID: 29772427.

Erickson PS, Kalscheur KF. "Nutrition and feeding of dairy cattle." Animal Agriculture. 2020:157–80. doi: 10.1016/B978-0-12-817052-6.00009-4. Epub 2020 Jan 24. PMCID: PMC7153313.

Eskenazi B, Gunier RB, Rauch S, et al. "Association of Lifetime Exposure to Glyphosate and Aminomethylphosphonic Acid (AMPA) with Liver Inflammation and Metabolic Syndrome at Young Adulthood: Findings from the CHAMACOS Study." Environmental Health Perspectives. 2023 Mar;131(3):37001. DOI: 10.1289/ehp11721. PMID: 36856429; PMCID: PMC9976611.

Essington, Dr. Michael E., "Class Notes," Soil Chemistry, University of Tennessee https://Web.utk.edu/~drtd0c/SoilErosion.pdf

Evans, Sam, "ISSUES 1990'S: COMMODITY PROGRAMS Acreage Reduction Programs", USDA, ERS, Bulletin #664-44,

1993, https://naldc.nal.usda.gov/download/CAT30948862/PDF

Evonik, "Unleash your livestock's full potential with amino acids", All About Feed, 2022, https://www.allaboutfeed.net/animal-feed/feed-additives/unleash-your-livestocks-full-potential-with-amino-acids/

FAO, "2022 Global Report on Food Crisis", 2022, https://www.fao.org/3/cb9997en/cb9997en.pdf

FAO, "Fertilizer Use to Surpass 200 Million Tonnes in 2018" 16 February, 2015, https://www.fao.org/news/story/en/item/277488/icode/

FAO, "Status of the World's Soil Resources Main Report", FAO 2015 ISBN 978-92-5-109004-6

FAO, "World Food Situation", 02/06/2023, https://www.fao.org/worldfoodsituation/csdb/en/

Farm Bureau, "Changes on the Horizon for the CRP" https://www.fb.org/market-intel/changes-on-the-horizon-for-the-crp

Farooqui AA, Farooqui T, Panza F, Frisardi V. "Metabolic syndrome as a risk factor for neurological disorders." Cell Mol Life Sci. 2012 Mar;69(5):741-62. doi: 10.1007/s00018-011-0840-1. Epub 2011 Oct 15. PMID: 21997383.

Federal Reserve History, "The Great Depression", 2013, https://www.federalreservehistory.org/essays/great-depression

Feltracco M, Barbaro E, Maule F, Bortolini M, Gabrieli J, De Blasi F, Cairns WR, Dallo F, Zangrando R, Barbante C, Gambaro A. "Airborne polar pesticides in rural and mountain sites of North-Eastern Italy: An emerging air quality issue." Environ Pollut. 2022 Sep 1;308:119657. doi: 10.1016/j.envpol.2022.119657. Epub 2022 Jun 21. PMID: 35750305.

Fernandes, Gracieli & Aparicio, Virginia & De Gerónimo, Eduardo & Bastos, Marília & Labanowski, Jérôme & Prestes, Osmar & Zanella, Renato & dos Santos, Danilo. (2018). "Indiscriminate use of glyphosate impregnates river epilithic biofilms in southern Brazil." Science of The Total Environment. 651. 10.1016/j.scitotenv.2018.09.292.

Ferrell, J. A., Dittmar, P. J., Sellers, B. A., and Devkota, P., " Herbicide Residues in Manure, Compost, or Hay", University of Florida, IFAS, 8/10/2020, https://edis.ifas.ufl.edu/publication/AG416

Fields S. "Global nitrogen: cycling out of control." Environ Health Perspect. 2004 Jul;112(10):A556-63. doi: 10.1289/ehp.112-a556. PMID: 15238298; PMCID: PMC1247398.

Fisher, Art, Walker, Mark, Powell, Pam, "DDT and DDE:Sources of Exposure and How to Avoid Them" University of Nevada, Reno, Extension Education, SP-03-16 NAES #52031334, 2023, https://ag.arizona.edu/region9wq/pdf/nv_walkerddtdde.pdf

FMC Agricultural Solutions, "Accolade", 6/24/17, https://fmccrop.grower/Products/Herbicides/Accolade.aspx

Fog, Kåre. (2008). "The effect of added nitrogen on the rate of decomposition of organic matter." Biological Reviews. 63. 433 - 462. 10.1111/j.1469-185X.1988.tb00725.x.

Foley, M., Nafziger, E., Slife, F., & Wax, L. (1983). "Effect of Glyphosate on Protein and Nucleic Acid Synthesis and ATP Levels in Common Cocklebur (Xanthium pensylvanicum) Root Tissue." Weed Science, 31(1), 76-80. doi:10.1017/S0043174500068570

Forlani, G., Mangiagalli, A., Nielsen, E., Suardi, C.M., "Degradation of the phosphonate herbicide glyphosate in soil: evidence for a possible involvement of unculturable microorganisms," Soil Biology and Biochemistry, Vol. 31, Issue 7,1999, Pages 991-997, https://doi.org/10.1016/S0038-0717(99)00010-3.

Friedman M. "Analysis, Nutrition, and Health Benefits of Tryptophan." Int J Tryptophan Res. 2018 Sep 26;11:1178646918802282. doi: 10.1177/1178646918802282. PMID: 30275700; PMCID: PMC6158605.

Gao A, Kouznetsova VL, Tsigelny IF. "Bovine leukemia virus relation to human breast cancer: Meta-analysis." Microb Pathog. 2020 Dec;149:104417. doi: 10.1016/j.micpath.2020.104417. Epub 2020 Jul 27. PMID: 32731009; PMCID: PMC7384413.

Garcia-Ruiz, Jose M., Begueria, Santiago, Nadal-Romero, Estela, Gonzalez-Hidalgo, Jose C., Lana-Renault, Noemi, Sanjuan ,

Yasmina, "A meta-analysis of soil erosion rates across the world,"Geomorphology, Volume 239, 2015,P. 160-173 https://doi.org/10.1016/j.geomorph.2015.03.008

Gassmann AJ, Petzold-Maxwell JL, Keweshan RS, Dunbar MW. "Field-evolved resistance to Bt maize by western corn rootworm." PLoS One. 2011;6(7):e22629. doi: 10.1371/journal.pone.0022629. Epub 2011 Jul 29. PMID: 21829470; PMCID: PMC3146474.

Gazey, C., Abbott, L., & Robson, A. (2004). "Indigenous and introduced arbuscular mycorrhizal fungi contribute to plant growth in two agricultural soils from south-western Australia" Mycorrhiza, 14(6), 355-362. https://doi.org/10.1007/s00572-003-0282-1

Gdanetz, Kristi and Trail, Frances, "The Wheat Microbiome Under Four Management Strategies, and Potential for Endophytes in Disease Protection" Phytobiomes Journal, 25 Oct 2017, https://apsjournals.apsnet.org/doi/full/10.1094/pbiomes-05-17-0023-r

GEPNER, J., HALL, L. & SATTELLE, D. "Insect acetylcholine receptors as a site of insecticide action." Nature 276, 188–190 (1978). https://doi.org/10.1038/276188a0

Giaccio GCM, Saez JM, Estévez MC, Salinas B, Corral RA, De Gerónimo E, Aparicio V, Álvarez A. "Developing a glyphosate-bioremediation strategy using plants and actinobacteria: Potential improvement of a riparian environment." J Hazard Mater. 2023 Mar 15;446:130675. doi: 10.1016/j.jhazmat.2022.130675. Epub 2022 Dec 25. PMID: 36608579.

Gietzen DW, Aja SM. "The brain's response to an essential amino acid-deficient diet and the circuitous route to a better meal." Mol Neurobiol. 2012 Oct;46(2):332-48. doi: 10.1007/s12035-012-8283-8. Epub 2012 Jun 7. PMID: 22674217; PMCID: PMC3469761.

Gilbert C. Fite, "Agricultural Adjustment Act," The Encyclopedia of Oklahoma History and Culture, https://www.okhistory.org/publications/enc/entry.php?entry=AG002.

Gill, Cassandra, "The transformation of food in America in the 19th century", Oxford University Press, October 7th 2016, https://blog.oup.com/2016/10/immigrants-food-america/

Gonçalves, Ana & Rocha, Carolina & Marques, João & Gonçalves, Fernando. (2021). "Fatty acids as suitable biomarkers to assess pesticide impacts in freshwater biological scales -A review." Ecological Indicators. 122. 1470-160. 10.1016/j.ecolind.2020.107299.

Grant, Daniel, "Port of New Orleans critical to growth of ag exports", Farm Week Now, Mar 11, 2022, https://www.farmweeknow.com/profitability/port-of-new-orleans-critical-to-growth-of-ag-exports/article_c5876580-a098-11ec-b586-d3f61616db1d.html

Greenblatt, James M., "Amino Acids: Why Don't I Have Enough?", Psychology Today, June 7, 2014, https://www.psychologytoday.com/us/blog/answers-appetite/201406/amino-acids-why-don-t-i-have-enough

Gresner, N., Rodehutscord, M., & Südekum, K.-H. (2022). "Amino acid pattern of rumen microorganisms in cattle fed mixed diets—An update." Journal of Animal Physiology and Animal Nutrition, 1, 752– 771. https://doi.org/10.1111/jpn.13676

GRO Intelligence, "World Wheat Reserves Outside of China to Drop to 14-Year Low", 2022, https://www.gro-intelligence.com/insights/world-wheat-reserves-to-drop-to-14-year-low

Gunders, Dana, "Wasted: How America Is Losing Up to 40 Percent of Its Food from Farm to Fork to Landfill", NRDC Issue Paper, Aug. 2012 IP:12-06-B https://www.nrdc.org/sites/default/files/wasted-food-IP.pdf

Guo C, Yang Y, Shi MX, Wang B, Liu JJ, Xu DX, Meng XH. "Critical time window of fenvalerate-induced fetal intrauterine growth restriction in mice." Ecotoxicol Environ Saf. 2019 May 15;172:186-193. doi: 10.1016/j.ecoenv.2019.01.054. Epub 2019 Jan 30. PMID: 30708230.

Hagervall TG, Jönsson YH, Edmonds CG, McCloskey JA, Björk GR. "Chorismic acid, a key metabolite in modification of tRNA." J Bacteriol. 1990 Jan;172(1):252-9. doi:

10.1128/jb.172.1.252-259.1990. PMID: 2104604; PMCID: PMC208425.

Hansen, Zeynep K., and Gary D. Libecap. "Small Farms, Externalities, and the Dust Bowl of the 1930s." Journal of Political Economy 112, no. 3 (2004): 665–94. https://doi.org/10.1086/383102.

Harpreet Kaur, Madhurama Gangwar and Anu Kalia (2015); Diversity of Actinomycetes from fodder leguminous plants and their biocontrol potential Int. J. of Adv. Res. 3 (Aug). 1141-1151 (ISSN 2320-5407). www.journalijar.com

Hashemy-Tonkabony, S & Mosstofian, B. (1979). "Chlorinated Pesticide Residues in Chicken Egg. Poultry science." 58. 1432-4. 10.3382/ps.0581432.

Hayes TB, Collins A, Lee M, Mendoza M, Noriega N, Stuart AA, Vonk A. "Hermaphroditic, demasculinized frogs after exposure to the herbicide atrazine at low ecologically relevant doses." Proc Natl Acad Sci U S A. 2002 Apr 16;99(8):5476-80. doi: 10.1073/pnas.082121499. PMID: 11960004; PMCID: PMC122794.

Hayes, Jared, "Federal Lawmakers Harvest $15 Million in Farm Subsidies", Environmental Working Group, 2017, https://www.ewg.org/news-insights/news/federal-lawmakers-harvest-15-million-farm-subsidies

He X, Tu Y, Song Y, Yang G, You M. "The relationship between pesticide exposure during critical neurodevelopment and autism spectrum disorder: A narrative review." Environ Res. 2022 Jan;203:111902. doi: 10.1016/j.envres.2021.111902. Epub 2021 Aug 17. PMID: 34416252.

Heidbüchel K, Raabe J, Baldinger L, Hagmüller W, Bussemas R. "One Iron Injection Is Not Enough-Iron Status and Growth of Suckling Piglets on an Organic Farm." Animals (Basel). 2019 Sep 4;9(9):651. doi: 10.3390/ani9090651. PMID: 31487865; PMCID: PMC6770926.

Herman, Jody L., Flores, Andrew R., Brown, Taylor N.T., Wilson, Bianca D.M., and Conron, Kerith J. "Age of Individuals Who Identify As Transgender in the United States", The Williams Institute, UCLA, School of Law, 2017, http://williamsinstitute.law.ucla.edu/wp-content/uploads/Age-Trans-Individuals-Jan-2017.pdf

Herman, Jody L., Flores, Andrew R., O'Neill, Kathryn K. ,"How Many Adults and Youth Identify as Transgender in the United States?", UCLA, School of Law, Williams Institute, 2022, https://williamsinstitute.law.ucla.edu/publications/trans-adults-united-states/

Hibbard, B. H. (1925). "Review of The Agricultural Crisis, 1920-23., by R. R. Enfield". Political Science Quarterly, 40(3), 476–478. https://doi.org/10.2307/2142221

Ho, Mae-Wan, "Glyphosate/Roundup & Human Male Infertility", Permaculture Research Institute, 2014, https://www.permaculturenews.org/2014/03/21/glyphosate-roundup-human-male-infertility/

Hribar, Carrie, "Understanding Concentrated Animal Feeding Operations and Their Impact on Communities" National Association of Local Boards of Health, 2010, https://www.cdc.gov/nceh/ehs/docs/understanding_cafos_nalboh.pdf

Humphries, Dave, Byrtus, Gary , and Anderson, Anne-Marie, "Glyphosate Residues In Alberta's Atmospheric Deposition, Soils And Surface Waters" Environmental Monitoring and Evaluation Branch Alberta Environment, 2005, https://open.alberta.ca/dataset/a4381736-cd17-4be1-b8ed-16aee8073be9/resource/5744d27f-fce1-43fd-a109-8ec6423929b4/download/6444.pdf

Illinois Department of Public Health, "Methyl Parathion", 2023, https://dph.illinois.gov/topics-services/diseases-and-conditions/diseases-a-z-list/methyl-parathion.html,

Ingaramo P, Alarcón R, Muñoz-de-Toro M, Luque EH. "Are glyphosate and glyphosate-based herbicides endocrine disruptors that alter female fertility?" Mol Cell Endocrinol. 2020 Dec 1;518:110934. doi: 10.1016/j.mce.2020.110934. Epub 2020 Jul 10. PMID: 32659439.

Ingham, Elaine, "Soil Food Web Price List" Soil Food Web Institute, 2023, https://www.soilfoodweb.com.au/our-services/price-list

Iummato MM, Fassiano A, Graziano M, Dos Santos Afonso M, Ríos de Molina MDC, Juárez ÁB. "Effect of glyphosate on the growth, morphology, ultrastructure and metabolism of Scenedesmus vacuolatus." Ecotoxicol Environ Saf. 2019 May

15;172:471-479. doi: 10.1016/j.ecoenv.2019.01.083. Epub 2019 Feb 6. PMID: 30738229.

Ivantsova E, Wengrovitz AS, Souders CL 2nd, Martyniuk CJ. "Developmental and behavioral toxicity assessment of glyphosate and its main metabolite aminomethylphosphonic acid (AMPA) in zebrafish embryos/larvae." Environ Toxicol Pharmacol. 2022 Jul;93:103873. doi: 10.1016/j.etap.2022.103873. Epub 2022 Apr 30. PMID: 35504511.

J. Mugnier, B. Mosse, "Spore germination and viability of a vesicular arbuscular mycorrhizal fungus, Glomus mosseae," Transactions of the British Mycological Society, Vol. 88, Issue 3, 1987, Pgs, 411-413, https://doi.org/10.1016/S0007-1536(87)80018-9.

Jackson E, Shoemaker R, Larian N, Cassis L. "Adipose Tissue as a Site of Toxin Accumulation." Compr Physiol. 2017 Sep 12;7(4):1085-1135. doi: 10.1002/cphy.c160038. Erratum in: Compr Physiol. 2018 Jun 18;8(3):1251. PMID: 28915320; PMCID: PMC6101675.

Jackson, James, "Weed Management with Manure in Mind", Texas A&M, Agrilife Extension, 2019, https://texasdairymatters.tamu.edu/files/2019/01/Weed-Management-with-Manure-in-Mind.pdf

Jackson, William R., "Humic, Fulvic, and Microbial Balance: Organic Soil Conditioning" Johnson Printing, 1993

Jbaily, A., Zhou, X., Liu, J. et al. "Air pollution exposure disparities across US population and income groups." Nature 601, 228–233 (2022). https://doi.org/10.1038/s41586-021-04190-y

Jefferies, Danica, "A potentially cancer-causing chemical is sprayed on much of America's farmland. Here is where it is used the most." NBC, 2022, https://www.nbcnews.com/data-graphics/toxic-herbicides-map-showing-high-use-state-rcna50052

Jensen PR, Gontang E, Mafnas C, Mincer TJ, Fenical W. Culturable marine actinomycete diversity from tropical Pacific Ocean sediments. Environ Microbiol. 2005 Jul;7(7):1039-48. doi: 10.1111/j.1462-2920.2005.00785.x. PMID: 15946301.

Johnson NC. "Can Fertilization of Soil Select Less Mutualistic Mycorrhizae?" Ecol Appl. 1993 Nov;3(4):749-757. doi: 10.2307/1942106. PMID: 27759303.

Johnson, Nancy C., and Gehring, Catherine A., "Mycorrhizas: Symbiotic Mediators of Rhizosphere and Ecosystem Processes," The Rhizosphere, Academic Press, 2007, P. 73-100, https://doi.org/10.1016/B978-012088775-0/50006-9.

Jones, H. Claire, "Attack of the Superweeds: Herbicides are losing the war — and agriculture might never be the same again.", The New York Times, 2021, https://www.nytimes.com/2021/08/18/magazine/superweeds-monsanto.html

Jose PA, Jha B. New Dimensions of Research on Actinomycetes: Quest for Next Generation Antibiotics. Front Microbiol. 2016 Aug 19;7:1295. doi: 10.3389/fmicb.2016.01295. PMID: 27594853; PMCID: PMC4990552.

Juge C, Samson J, Bastien C, Vierheilig H, Coughlan A, Piché Y. "Breaking dormancy is spores of the arbuscular mycorrhizal fungus Glomus intraradices: a critical cold-storage period." Mycorrhiza. 2002 Feb;12(1):37-42. doi: 10.1007/s00572-001-0151-8. PMID: 11968945.

Junior, Valdir & Lopes, Fernanda & Schwab, Charles & Toledo, Mateus & Collao, Edgar. (2021). "Effects of rumen-protected methionine supplementation on the performance of high production dairy cows in the tropics." PloS one. 16. e0243953. 10.1371/journal.pone.0243953.

Karagiannidis, N., Velemis, D., Stavropoulos, N., "Root colonization and spore population by VA-mycorrhizal fungi in four grapevine rootstocks", Journal of Grapevine Research, 2015-08-06, https://ojs.openagrar.de/index.php/VITIS/article/view/4856

Keating, Austin, "Illinois Facing Most Severe Erosion In Two Decades" Illinois Public Media, June 06, 2016, https://will.illinois.edu/news/story/illinois-facing-most-severe-erosion-in-two-decades

Kelleher, Brian P., and Simpson, Andre. J., "Humic Substances in Soils: Are They Really Chemically Distinct?"Environmental

Science & Technology 2006 40 (15), 4605-4611 https://doi.org/10.1021/es0608085

Kendall, Carol, "Resources on Isotopes", USGS, 1998, https://wwwrcamnl.wr.usgs.gov/isoig/period/n_iig.html

Kennedy, Pagan, "How to Get High on Soil", The Atlantic, 2012, https://www.theatlantic.com/health/archive/2012/01/how-to-get-high-on-soil/251935/

Kershner, Isabel, "Pollen Study Points to Drought as Culprit in Bronze Age Mystery", The New York Times, Oct. 22nd, 2013, https://www.nytimes.com/2013/10/23/world/middleeast/pollen-study-points-to-culprit-in-bronze-era-mystery.html

Khan, Asif & Sharif, Muhammad & Ali, Amjad & Shah, Syed Noor Muhammad & Mian, Ishaq Ahmad & Wahid, Fazli & Jan, Bismillah & Adnan, Muhammad & Nawaz, Shah & Ali, Nisar. (2014). "Potential of AM Fungi in Phytoremediation of Heavy Metals and Effect on Yield of Wheat Crop." American Journal of Plant Sciences. 5. 1578-1586. 10.4236/ajps.2014.511171.

Khan, Z., Abubakar, M., Arshed, M.J. et al. "Molecular investigation of possible relationships concerning bovine leukemia virus and breast cancer." Sci Rep 12, 4161 (2022). https://doi.org/10.1038/s41598-022-08181-5

Kill, Carina, "Bug Off: The Neural Effects of Insecticides", Grey Matters, The Undergraduate Neuroscience Journal, 2019, https://greymattersjournal.org/bug-off-the-neural-effects-of-insecticides/

King, F. H. (Franklin Hiram), "A text book of the physics of agriculture", Madison, Wis, 1907

Kinoshita, K., Otsuka, R., Takada, M. et al. "Low Amino Acid Score of Breakfast is Associated with the Incidence of Cognitive Impairment in Older Japanese Adults: A Community-Based Longitudinal Study." J Prev Alzheimers Dis 9, 151–157 (2022). https://doi.org/10.14283/jpad.2021.25

Kirkby, Clive A., Richardson, Alon E., Wade, Len, Passioura, John B., Batten, Graeme D., Blanchard, Chris, Kirkegaard, John A., "Nutrient availability limits carbon sequestration in arable soils", Soil Biology and Biochemistry, Vol.68, 2014, Pages 402-409, https://doi.org/10.1016/j.soilbio.2013.09.032.

Klein, Christopher, "10 Things You May Not Know About the Dust Bowl", The History Channel, August 24, 2012, https://www.history.com/news/10-things-you-may-not-know-about-the-dust-bowl

Koide, R.T. and Dickie, I.A., "Effects of Mycorrhizae Fungus on Plant Populations", Plant & Soil, p. 307-317, 2002

KOIZUMI,Iwao, SUZUKI,Yoshihiko, KANEKO,J. J., "Studies on the Fatty Acid Composition of Intramuscular Lipids of Cattle, Pigs and Birds", Journal of Nutritional Science and Vitaminology, 1991 Volume 37, https://doi.org/10.3177/jnsv.37.545

Kovak, Emma and Rejto, Dan Blaustein, "The World's First Genetically Engineered Wheat Is Here", The Breakthrough Institute, 2022, https://thebreakthrough.org/issues/food-agriculture-environment/the-worlds-first-genetically-engineered-wheat-is-here

Krishnamurthy, A., Moore, J. K., Zender, C. S., and Luo, C. (2007), "Effects of atmospheric inorganic nitrogen deposition on ocean biogeochemistry," J. Geophys. Res., 112, G02019, doi:10.1029/2006JG000334.

Krueger, Monika & Schledorn, Philipp & Schrödl, Wieland & Hoppe, Hans-Wolfgang & Lutz, Walburga & Shehata, Awad. (2014). "Detection of glyphosate residues in animals and humans." Environmental & analytical toxicology. 4. 10.4172/2161-0525.1000210.

Kuntz , Marcel, "Transgenic Plants and Beyond", Advances in Botanical Research, February 14, 2018

Kwiatkowska M, Jarosiewicz P, Michałowicz J, Koter-Michalak M, Huras B, Bukowska B. "The Impact of Glyphosate, Its Metabolites and Impurities on Viability, ATP Level and Morphological changes in Human Peripheral Blood Mononuclear Cells." PLoS One. 2016 Jun 9;11(6):e0156946. doi: 10.1371/journal.pone.0156946. PMID: 27280764; PMCID: PMC4900596.

Lam, Hon-Ming & Coschigano, K.T. & Oliveira, I.C. & Melo-Oliveira, R. & Coruzzi, Gloria. (1996). "The Molecular-Genetics of Nitrogen Assimilation into Amino Acids in Higher Plants. Annual review of plant physiology and plant molecular biology." 47. 569-593. 10.1146/annurev.arplant.47.1.569.

Lauer, Franziska & Pätzold, Stefan & Gerlach, Renate & Protze, Jens & Willbold, S. & Amelung, Wulf. (2013). "Phosphorus status

in archaeological arable topsoil relicts—Is it possible to reconstruct conditions for prehistoric agriculture in Germany?". Geoderma. s 207–208. 111–120. 10.1016/j.geoderma.2013.05.005.
- Lehmann J, Kleber M. "The contentious nature of soil organic matter." Nature. 2015 Dec 3;528(7580):60-8. doi: 10.1038/nature16069. Epub 2015 Nov 23. PMID: 26595271.
- Lenox, Catherine, "Role of Dietary Fatty Acids in Dogs & Cats", Today's Veterinary Practice, August 19, 2016, https://todaysveterinarypractice.com/nutrition/role-of-dietary-fatty-acids-in-dogs-cats/
- Li, Jiemei & Zhu, Tianjie & Wang, F & Qiu, Xinghua & Lin, Weili. (2006). "Observation of organochlorine pesticides in the air of the Mt. Everest region." Ecotoxicology and environmental safety. 63. 33-41. 10.1016/j.ecoenv.2005.04.001.
- Li, Jun & Zhang, Gan & Guo, Lingli & Xu, Weihai & Li, Xiang-Dong & Lee, Celine Siu Lan & Ding, Aijun & Wang, Tao. (2007). "Organochlorine Pesticides in the Atmosphere of Guangzhou and Hong Kong: Regional Sources and Long-Range Atmospheric Transport." Atmospheric Environment. 41. 3889-3903. 10.1016/j.atmosenv.2006.12.052.
- Li, Y., Lohmann, R., Zou, X., Wang, C. and Zhang, L., 2020. "Air-water exchange and distribution pattern of organochlorine pesticides in the atmosphere and surface water of the open Pacific ocean." Environmental Pollution, 265, p.114956.
- Lincicome, Scott, "Examining America's Farm Subsidy Problem", CATO Institute, 2020, https://www.cato.org/commentary/examining-americas-farm-subsidy-problem#
- Livingston, Gretchen, "Is U.S. fertility at an all-time low? Two of three measures point to yes", Pew Research Center, 2019, https://www.pewresearch.org/short-reads/2019/05/22/u-s-fertility-rate-explained/
- Lobe, I., Amelung, W. and Du Preez, C.C. (2001), "Losses of carbon and nitrogen with prolonged arable cropping from sandy soils of the South African Highveld." European Journal of Soil Science, 52: 93-101. https://doi.org/10.1046/j.1365-2389.2001.t01-1-00362.x

Lockett, Eleesha, "Grounding: Exploring Earthing Science and the Benefits Behind It", Healthline, 2023, https://www.healthline.com/health/grounding

Lotterman, Edward, "The porcine slaughter of the innocents" Federal Reserve Bank of Minneapolis, 1999, https://www.minneapolisfed.org/article/1999/the-porcine-slaughter-of-the-innocents

Lu W, Li L, Chen M, Zhou Z, Zhang W, Ping S, Yan Y, Wang J, Lin M. "Genome-wide transcriptional responses of Escherichia coli to glyphosate, a potent inhibitor of the shikimate pathway enzyme 5-enolpyruvylshikimate-3-phosphate synthase." Mol Biosyst. 2013 Mar;9(3):522-30. doi: 10.1039/c2mb25374g. Epub 2012 Dec 18. PMID: 23247721.

Lushchak VI, Matviishyn TM, Husak VV, Storey JM, Storey KB. "Pesticide toxicity: a mechanistic approach." EXCLI J. 2018 Nov 8;17:1101-1136. doi: 10.17179/excli2018-1710. PMID: 30564086; PMCID: PMC6295629.

Lyu CP, Pei JR, Beseler LC, Li YL, Li JH, Ren M, Stallones L, Ren SP. "Case Control Study of Impulsivity, Aggression, Pesticide Exposure and Suicide Attempts Using Pesticides among Farmers." Biomed Environ Sci. 2018 Mar;31(3):242-246. doi: 10.3967/bes2018.031. PMID: 29673448; PMCID: PMC7413211.

Maeda H, Dudareva N. "The shikimate pathway and aromatic amino Acid biosynthesis in plants." Annu Rev Plant Biol. 2012;63:73-105. doi: 10.1146/annurev-arplant-042811-105439. PMID: 22554242.

Main, Douglas, "Glyphosate Now the Most Used Agricultural Chemical Ever", Newsweek Tech and Science, 2/2/16, https://www.newsweek.com/glyphosate-now-most-used-agricultural-chemical-ever-422419

Mamontova EA, Mamontov AA. "Spatial and Temporal Variations of Polychlorinated Biphenyls and Organochlorine Pesticides in Snow in Eastern Siberia." Atmosphere. 2022; 13(12):2117. https://doi.org/10.3390/atmos13122117

Mantri, S., Fullard, M.E., Beck, J. et al. "State-level prevalence, health service use, and spending vary widely among Medicare beneficiaries with Parkinson disease." npj Parkinson's Disease 5, 1 (2019). https://doi.org/10.1038/s41531-019-0074-8

Marshall, Lisa, "Why dirt may be nature's original stress-buster", University of Colorado, Boulder, 2018, https://www.colorado.edu/today/2019/05/09/natures-original-stress-buster

Marx, Karl, "Communist Manifesto", Progress Publishers, Moscow, 1969, https://www.marxists.org/archive/marx/works/1848/communist-manifesto/ch02.htm

Masayuki Hayakawa, Hideo Nonomura, Humic acid-vitamin agar, a new medium for the selective isolation of soil actinomycetes, Journal of Fermentation Technology, Vol. 65, Issue 5, 1987, Pgs. 501-509, https://doi.org/10.1016/0385-6380(87)90108-7.

Mattos, Ricardo, Staples, Charles R., Thatcher, William W., "Effects of Dietary Acids on Reproduction in Ruminants", Journals of Reproduction and Fertility, 2000, http://virtusnutrition.com/wp-content/uploads/2016/11/Fatty-Acids-and-Reproduction-Mattos-et.-al..pdf

McFarling, Usha Lee, "On the Texas-Mexico border, a bold plan to diversify Alzheimer's research takes shape", STAT, 2022, https://www.statnews.com/2022/09/28/bold-plan-diversify-alzheimers-research-takes-shape/

McKillen, Elizabeth, "The Socialist Party of America, 1900–1929", Oxford Research Encyclopedia, 2017, https://doi.org/10.1093/acrefore/9780199329175.013.413

McVan, Madison, "18 years and counting: EPA still has no method for measuring CAFO air pollution", Williston Herald Media, 2023, https://www.willistonherald.com/news/farm_and_ranch/18-years-and-counting-epa-still-has-no-method-for-measuring-cafo-air-pollution/article_5b1c9e3c-e083-11ed-9e47-1fae224891af.html

Mehrotra S, Goyal V. Agrobacterium-mediated gene transfer in plants and biosafety considerations. Appl Biochem Biotechnol. 2012 Dec;168(7):1953-75. doi: 10.1007/s12010-012-9910-6. Epub 2012 Oct 23. PMID: 23090683.

Melgar MJ, Santaeufemia M, Garcia MA. "Organophosphorus pesticide residues in raw milk and infant formulas from Spanish northwest." J Environ Sci Health B. 2010 Oct;45(7):595-600. doi: 10.1080/03601234.2010.502394. PMID: 20803361.

Mendelsohn, Mike, Kough, John, Vaituzis, Zigfridais ,and Matthews, Keith, "Are Bt crops safe?" Nature Biotechnology, 2003, https://19january2021snapshot.epa.gov/sites/static/files/2015-08/documents/are_bt_crops_safe.pdf

Menzel, Ralph, Geweiler, Diana, Sass, Annika, Simsek, Dilara, Ruess, Liliane, "Nematodes as Important Source for Omega-3 Long-Chain Fatty Acids in the Soil Food Web and the Impact in Nutrition for Higher Trophic Levels" Frontiers in Ecology and Evolution Vol.6, 2018 DOI=10.3389/fevo.2018.00096 https://www.frontiersin.org/articles/10.3389/fevo.2018.00096

Mercurio P, Flores F, Mueller JF, Carter S, Negri AP. "Glyphosate persistence in seawater." Mar Pollut Bull. 2014 Aug 30;85(2):385-90. doi: 10.1016/j.marpolbul.2014.01.021. Epub 2014 Jan 24. PMID: 24467857.

Middleton N, Kang U. "Sand and Dust Storms: Impact Mitigation." Sustainability. 2017; 9(6):1053. https://doi.org/10.3390/su9061053

Milesi MM, Lorenz V, Pacini G, Repetti MR, Demonte LD, Varayoud J, Luque EH. "Perinatal exposure to a glyphosate-based herbicide impairs female reproductive outcomes and induces second-generation adverse effects in Wistar rats." Arch Toxicol. 2018 Aug;92(8):2629-2643. doi: 10.1007/s00204-018-2236-6. Epub 2018 Jun 9. PMID: 29947892.

Miller N, Estoup A, Toepfer S, Bourguet D, Lapchin L, Derridj S, Kim KS, Reynaud P, Furlan L, Guillemaud T. "Multiple transatlantic introductions of the western corn rootworm." Science. 2005 Nov 11;310(5750):992. doi: 10.1126/science.1115871. PMID: 16284172.

Miller, Jenesse, "Study finds some of the world's lowest dementia rates in Amazonian indigenous groups", USC, University of Southern California, 2022, https://news.usc.edu/197541/some-of-the-worlds-lowest-dementia-rates-are-found-in-amazonian-indigenous-groups/

Mineur YS, Obayemi A, Wigestrand MB, Fote GM, Calarco CA, Li AM, Picciotto MR. "Cholinergic signaling in the hippocampus regulates social stress resilience and anxiety- and depression-like behavior." Proc Natl Acad Sci U S A. 2013 Feb 26;110(9):3573-8. doi: 10.1073/pnas.1219731110. Epub 2013 Feb 11. PMID: 23401542; PMCID: PMC3587265.

Miranda, R.A., Silva, B.S., de Moura, E.G. et al. "Pesticides as endocrine disruptors: programming for obesity and diabetes." Endocrine 79, 437–447 (2023). https://doi.org/10.1007/s12020-022-03229-y

Mladenović M, Arsić BB, Stanković N, Mihović N, Ragno R, Regan A, Milićević JS, Trtić-Petrović TM, Micić R. "The Targeted Pesticides as Acetylcholinesterase Inhibitors: Comprehensive Cross-Organism Molecular Modelling Studies Performed to Anticipate the Pharmacology of Harmfulness to Humans In Vitro." Molecules. 2018 Aug 30;23(9):2192. doi: 10.3390/molecules23092192. PMID: 30200244; PMCID: PMC6225315.

Mnif W, Hassine AI, Bouaziz A, Bartegi A, Thomas O, Roig B. "Effect of endocrine disruptor pesticides: a review." Int J Environ Res Public Health. 2011 Jun;8(6):2265-303. doi: 10.3390/ijerph8062265. Epub 2011 Jun 17. PMID: 21776230; PMCID: PMC3138025.

Mohammad Abass Ahanger, Abeer Hashem, Elsayed Fathi Abd-Allah, Parvaiz Ahmad, "Arbuscular Mycorrhiza in Crop Improvement under Environmental Stress," Emerging Technologies and Management of Crop Stress Tolerance, Academic Press, 2014, Pages 69-95, https://doi.org/10.1016/B978-0-12-800875-1.00003-X.

Montoya AL, Chilton MD, Gordon MP, Sciaky D, Nester EW. Octopine and nopaline metabolism in Agrobacterium tumefaciens and crown gall tumor cells: role of plasmid genes. J Bacteriol. 1977 Jan;129(1):101-7. doi: 10.1128/jb.129.1.101-107.1977. PMID: 830636; PMCID: PMC234901.

Moore, G.," Plowing Under Cotton and Killing Pigs", The Friday Footnote, NCSU, 2020, https://footnote.wordpress.ncsu.edu/2020/08/14/plowing-under-cotton-and-killing-pigs-8-14-2020/

Mooshammer, Maria & Wanek, Wolfgang & Hämmerle, Ieda & Fuchslueger, Lucia & Hofhansl, Florian & Knoltsch, Anna & Schnecker, Jörg & Takriti, Mounir & Watzka, Margarete & Wild, Birgit & Keiblinger, Katharina & Zechmeister-Boltenstern, Sophie & Richter, Andreas. (2014). "Adjustment of microbial nitrogen use efficiency to carbon:nitrogen imbalances regulates soil nitrogen cycling." Nature communications. 5. 3694. 10.1038/ncomms4694.

Muhs, D. R., Budahn, J. R., Prospero, J. M., and Carey, S. N. (2007), "Geochemical evidence for African dust inputs to soils of western Atlantic islands: Barbados, the Bahamas, and Florida," J. Geophys. Res., 112, F02009, doi:10.1029/2005JF000445.

Mulvaney, R.L., Khan, S. A., and Ellsworth, T. R., "Synthetic Nitrogen Fertilizers Deplete Soil Nitrogen: A Global Dilemma for Sustainable Cereal", University of Illinois, 2009, https://acsess.onlinelibrary.wiley.com/doi/pdf/10.2134/jeq 2008.0527

Myers JP, Antoniou MN, Blumberg B, Carroll L, Colborn T, Everett LG, Hansen M, Landrigan PJ, Lanphear BP, Mesnage R, Vandenberg LN, Vom Saal FS, Welshons WV, Benbrook CM. "Concerns over use of glyphosate-based herbicides and risks associated with exposures: a consensus statement." Environ Health. 2016 Feb 17;15:19. doi: 10.1186/s12940-016-0117-0. PMID: 26883814; PMCID: PMC4756530.

Nafici, Saara, "Weed of the Month: Lambsquarters", Brooklyn Botanic Garden, 2018, https://www.bbg.org/article/weed_of_the_month_lambsqua rters

Nair AA, Yu F. "Quantification of Atmospheric Ammonia Concentrations: A Review of Its Measurement and Modeling." Atmosphere. 2020; 11(10):1092. https://doi.org/10.3390/atmos11101092

Nanalyze, "4 Companies Replacing Nitrogen in Fertilizers" 2021, https://www.nanalyze.com/2021/07/companies-replacing-nitrogen-fertilizers/

Näsholm, T., Ekblad, A., Nordin, A. et al. "Boreal forest plants take up organic nitrogen." Nature 392, 914–916 (1998). https://doi.org/10.1038/31921

National Farmers Union, "Price-Fixing Indictment Highlights Need for Stronger Antitrust Enforcement, Protections for Farmers", 2020, https://nfu.org/2020/06/03/price-fixing-indictment-highlights-need-for-stronger-antitrust-enforcement-protections-for-farmers/

National Minerals Information Center, "Nitrogen Statistics and Information", 2023, https://www.usgs.gov/centers/national-minerals-information-center/nitrogen-statistics-and-information

National Pesticide Information Center, "Bacillus thuringiensis (Bt)", 2023, http://npic.orst.edu/factsheets/btgen.html#symptoms

NatWestGroup Remembers, "Gold, banknotes and money supply in the First World War", 2022, https://www.natwestgroupremembers.com/banking-in-wartime/banking-business/gold-banknotes-and-money-supply-in-the-first-world-war.html

Naushad SM, Jain JM, Prasad CK, Naik U, Akella RR. "Autistic children exhibit distinct plasma amino acid profile." Indian J Biochem Biophys. 2013 Oct;50(5):474-8. PMID: 24772971.

Naz MSG, Tehrani FR, Majd HA, Ahmadi F, Ozgoli G, Fakari FR, Ghasemi V. "The prevalence of polycystic ovary syndrome in adolescents: A systematic review and meta-analysis." Int J Reprod Biomed. 2019 Sep 3;17(8):533-542. doi: 10.18502/ijrm.v17i8.4818. PMID: 31583370; PMCID: PMC6745085.

Niederhuber, Matthew, "Insecticidal Plants: The Tech and Safety of GM Bt Crops", Harvard University, Graduate School of Arts and Sciences, August 10, 2015 https://sitn.hms.harvard.edu/flash/2015/insecticidal-plants/

Nielsen, D.C. and Calderón, F.J. (2011). "Fallow Effects on Soil. In Soil Management: Building a Stable Base for Agriculture" (eds J.L. Hatfield and T.J. Sauer). https://doi.org/10.2136/2011.soilmanagement.c19

Nightingale, Sarah, "Research showing how nematodes use smell to select new insect hosts could improve biological control of crop pests" July 24th, 2017, https://phys.org/news/2017-07-nematodes-insect-hosts-biological-crop.html

Nikitina, E., Burk-Körner, A., Wiesenfarth, M., Alwers, E., Heide, D., Tessmer, C., Ernst, C., Krunic, D., Schrotz-King, P., Chang-

Claude, J., von Winterfeld, M., Herpel, E., Brobeil, A., Brenner, H., Heikenwalder, M., Hoffmeister, M., Kopp-Schneider, A. and Bund, T. (2023), "Bovine meat and milk factor protein expression in tumor-free mucosa of colorectal cancer patients coincides with macrophages and might interfere with patient survival." Mol Oncol. https://doi.org/10.1002/1878-0261.13390

Noemie, Cresto & Forner-Piquer, Isabel & Baig, Asma & Chatterjee, Mousumi & Perroy, Julie & Goracci, Jacopo & Marchi, Nicola. (2023). "Pesticides at brain borders: Impact on the blood-brain barrier, neuroinflammation, and neurological risk trajectories." Chemosphere. 324. 138251. 10.1016/j.chemosphere.2023.138251.

Noltemeyer, Matt, "Does GM wheat have a future in US?", World-Grain.com, 03.17.2023, https://www.world-grain.com/articles/18253-does-gm-wheat-have-future-in-us

North Carolina State University. "Field study shows how a GM crop can have diminishing success at fighting off insect pest." ScienceDaily. ScienceDaily, 21 May 2015.

Nurkanto, A., Lisdiyanti, P., Hamada, M. et al. "Actinoplanes bogoriensis sp. nov., a novel actinomycete isolated from leaf litter." J Antibiot 69, 26–30 (2016). https://doi.org/10.1038/ja.2015.81

Ojiro R, Okano H, Takahashi Y, Takashima K, Tang Q, Ozawa S, Zou X, Woo GH, Shibutani M. "Comparison of the effect of glyphosate and glyphosate-based herbicide on hippocampal neurogenesis after developmental exposure in rats." Toxicology. 2023 Jan 1;483:153369. doi: 10.1016/j.tox.2022.153369. Epub 2022 Nov 2. PMID: 36332718.

O'Neill Hayes, Tara, and Kerska, Katerina,"PRIMER: Agriculture Subsidies and Their Influence on the Composition of U.S. Food Supply and Consumption", American Action Forum, 2021, https://www.americanactionforum.org/research/primer-agriculture-subsidies-and-their-influence-on-the-composition-of-u-s-food-supply-and-consumption/

Orr, Richard, "Great Plains Soil Erosion Up 17 Pct." Chicago Tribune, April 03, 1985 https://www.chicagotribune.com/news/ct-xpm-1985-04-03-8501190047-story.html

Pace, Matthew, "Hidden Partners: Mycorrhizal Fungi and Plants" The New York Botanical Garden, 2003

Paddock, Catharine, "Soil Bacteria Work In Similar Way To Antidepressants", Medical News Today, 2007, https://www.medicalnewstoday.com/articles/66840#1

Pakkasmaa, Susanna, "PCB threatens polar bear populations", Aarhus University, 14 December 2016, https://arctic.au.dk/news-and-events/news/show/artikel/pcb-threatens-polar-bear-populations

Panseri, Sara & Nobile, Maria & Arioli, Francesco & Biolatti, Cristina & Pavlovic, Radmila & Chiesa, Luca. (2020). "Occurrence of perchlorate, chlorate and polar herbicides in different baby food commodities." Food Chemistry. 330. 127205. 10.1016/j.foodchem.2020.127205.

Pardío, Violeta & Ibarra, Nelly & Rodríguez, Miguel & Waliszewski, Krzysztof. (2001). "Use of Cholinesterase Activity in Monitoring Organophosphate Pesticide Exposure of Cattle Produced in Tropical Areas." Journal of agricultural and food chemistry. 49. 6057-62. 10.1021/jf010431g.

Parthasarathy A, Cross PJ, Dobson RCJ, Adams LE, Savka MA, Hudson AO. A "Three-Ring Circus: Metabolism of the Three Proteogenic Aromatic Amino Acids and Their Role in the Health of Plants and Animals." Front Mol Biosci. 2018 Apr 6;5:29. doi: 10.3389/fmolb.2018.00029. PMID: 29682508; PMCID: PMC5897657.

Paterson, Keith, "Where Was The Land Of Uz?", 2023, https://biblereadingarchaeology.com/about/

Pathogens and Manure https://www.extension.purdue.edu/extmedia/id/cafo/id-356.html

Peeples, Lynne, "Arsenic In Agriculture Enjoys Comeback In Poultry Feed, Pesticides", Huffington Post, Oct. 22nd, 2012, https://www.huffpost.com/entry/arsenics-industrial-agriculture-pesticides-poultry_n_2001340

Peirson, Erica, "Acetylcholine: How and Why to Optimize the Synthesis of this Vital Neurotransmitter", Pierson Center for Children, 2018, https://www.peirsoncenter.com/articles/acetylcholine-how-

and-why-to-optimize-the-synthesis-of-this-vital-neurotransmitter

Peixoto, Francisco. (2006). "Comparative effects of the Roundup and glyphosate on mitochondrial oxidative phosphorylation." Chemosphere. 61. 1115-22. 10.1016/j.chemosphere.2005.03.044.

Pellett, Peter L., "World essential amino acid supply with special attention to South-East Asia" Food and Nutrition Bulletin, vol. 17, no. 3 1996, The United Nations University. https://journals.sagepub.com/doi/pdf/10.1177/156482659601700304

Penn State Extension, "Avoiding Soil Compaction", February 12, 2005, https://extension.psu.edu/avoiding-soil-compaction

Penn State Extension, "Introduction to Weeds and Herbicides" 6/26/17/ https://www.extension.psu.edu/weeds/control/introduction-to-weeds-and-herbicides/herbicides

Perakis, S., Hedin, L. "Nitrogen loss from unpolluted South American forests mainly via dissolved organic compounds." Nature 415, 416–419 (2002). https://doi.org/10.1038/415416a

Peryea, F.J., Creger, T.L. "Vertical distribution of lead and arsenic in soils contaminated with lead arsenate pesticide residues."" Water Air Soil Pollut 78, 297–306 (1994). https://doi.org/10.1007/BF00483038

Peterson, Eric M., Green, Frank B., and Smith, Philip N. "Pesticides Used on Beef Cattle Feed Yards Are Aerially Transported into the Environment Via Particulate Matter" Environmental Science & Technology 2020 54 (20), DOI: 10.1021/acs.est.0c03603

Petrie JR, Shrestha P, Belide S, Mansour MP, Liu Q, Horne J, Nichols PD, Singh SP. "Transgenic production of arachidonic acid in oilseeds." Transgenic Res. 2012 Feb;21(1):139-47. doi: 10.1007/s11248-011-9517-7. Epub 2011 May 1. PMID: 21533900.

Pimentel, David & Burgess, Michael. (2013). "Soil Erosion Threatens Food Production." Agriculture. 3. 443-463. 10.3390/agriculture3030443.

Pipke R, Amrhein N. "Degradation of the Phosphonate Herbicide Glyphosate by Arthrobacter atrocyaneus" ATCC 13752. Appl Environ Microbiol. 1988 May;54(5):1293-6. doi: 10.1128/aem.54.5.1293-1296.1988. PMID: 16347639; PMCID: PMC202644.

Pipke R, Amrhein N. "Isolation and Characterization of a Mutant of Arthrobacter sp. Strain GLP-1 Which Utilizes the Herbicide Glyphosate as Its Sole Source of Phosphorus and Nitrogen." Appl Environ Microbiol. 1988 Nov;54(11):2868-70. doi: 10.1128/aem.54.11.2868-2870.1988. PMID: 16347784; PMCID: PMC204389.

Plumer, Brad, "No-Till Farming Is On The Rise. That's Actually A Big Deal." The Washington Post, Nov. 9th, 2013,

Poiger T, Buerge IJ, Bächli A, Müller MD, Balmer ME. "Occurrence of the herbicide glyphosate and its metabolite AMPA in surface waters in Switzerland determined with on-line solid phase extraction LC-MS/MS." Environ Sci Pollut Res Int. 2017 Jan;24(2):1588-1596. doi: 10.1007/s11356-016-7835-2. Epub 2016 Oct 27. PMID: 27787705.

Prakash N, Narayana K, Murthy GS, Moudgal NR, Honnegowda. "The effect of malathion, an organophosphate, on the plasma FSH, 17 beta-estradiol and progesterone concentrations and acetylcholinesterase activity and conception in dairy cattle." Veterinary and Human Toxicology. 1992 Apr;34(2):116-119. PMID: 1509669.

Pritchard J. Alberta. "Organophosphate toxicity in dairy cattle." Can Vet J. 1989 Feb;30(2):179. PMID: 17423244; PMCID: PMC1681043.

Pryor, F. (1985). "The Invention of the Plow. Comparative Studies in Society and History," 27(4), 727-743. doi:10.1017/S0010417500011749

Puffenbarger, Robyn A., "Molecular Biology of the Enzymes that Degrade Endocannabinoids" Current Drug Targets-CNS & Neurological Disorders, Volume 4, Number 6, 2005, pp. 625-631(7), Bentham Science Publishers, DOI: https://doi.org/10.2174/156800705774933050

Purdy, Michael C., "Parkinson's U.S. rates highest in whites, Hispanics, and Midwest, Northeast", Washington University in St. Louis, 2010, https://source.wustl.edu/2010/01/parkinson-us-rates-highest-in-whites-hispanics-and-midwest-northeast/

Qu, Chengkai & Doherty, Angela & Xing, Xinli & Sun, Wen & Albanese, Stefano & Lima, Annamaria & Qi, Shihua & De Vivo, Benedetto. (2017). "Polyurethane Foam-Based Passive Air Samplers in Monitoring Persistent Organic Pollutants:" Theory and Application. 10.1016/B978-0-444-63763-5.00021-5.

Ratcliff, Alice W. , Busse, Matt D., Shestak, Carol J., "Changes in microbial community structure following herbicide (glyphosate) additions to forest soils," Applied Soil Ecology,Vol. 34, Issues 2–3, 2006, P. 114-124, https://doi.org/10.1016/j.apsoil.2006.03.002.

Reddy NR, Flick GJ, Arganosa GC, Young RW. "Composition and Pesticide and Herbicide Residue Analysis of Fresh and 40-Year-Old Pasteurized Blue Crab (Callinectes sapidus) Meat." J Food Prot. 1991 Apr;54(4):298-301. doi: 10.4315/0362-028X-54.4.298. PMID: 31051625.

Reinhardt, Claudia and Ganzel, Bill, "Culling The Herds" Wessels, Living History Farm, 2003, https://livinghistoryfarm.org/farming-in-the-1930s/crops/culling-the-herds/

Reisig DD, Reay-Jones FP. Inhibition of Helicoverpa zea (Lepidoptera: Noctuidae) Growth by Transgenic Corn Expressing Bt Toxins and Development of Resistance to Cry1Ab. Environ Entomol. 2015 Aug;44(4):1275-85. doi: 10.1093/ee/nvv076. Epub 2015 May 20. PMID: 26314074.

Rillig, M.C., "Polymers & Microorganisms", Encyclopedia of the Soils in the Environment, 2005

Roberts, Owen, "UNITED STATES v. BUTLER et al.", Supreme Court, 1936

Rodríguez-Castellanos, L. and Sanchez-Hernandez, J.C. (2007), "Chemical reactivation and aging kinetics of organophosphorus-inhibited cholinesterases from two earthworm species." Environmental Toxicology and Chemistry, 26: 1992-2000. https://doi.org/10.1897/06-625R1.1

Rodríguez-Kábana R. "Organic and inorganic nitrogen amendments to soil as nematode suppressants." J Nematol. 1986 Apr;18(2):129-34. PMID: 19294153; PMCID: PMC2618534.

Rom Keshet, Ayelet Erez; Arginine and the metabolic regulation of nitric oxide synthesis in cancer. Dis Model Mech 1 August 2018; 11 (8): dmm033332. doi: https://doi.org/10.1242/dmm.033332

Ross, Mark and Edwards, Chris, "In New Zealand, Farmers Don't Want Subsidies", CATO Institute, 2012 https://www.cato.org/commentary/new-zealand-farmers-dont-want-subsidies#

Rossi M, Amaretti A, Raimondi S. "Folate production by probiotic bacteria." Nutrients. 2011 Jan;3(1):118-34. doi: 10.3390/nu3010118. Epub 2011 Jan 18. PMID: 22254078; PMCID: PMC3257725.

Rusch, A., Valantin-Morison, M., Sarthou, J. P., & Roger-Estrade, J. (2013). "Effect of crop management and landscape context on insect pest populations and crop damage." Agriculture, Ecosystems Environment, 166, 118–125. https://doi.org/10.1016/J.AGEE.2011.05.004

Sabio Y García CA, Schiaffino MR, Lozano VL, Vera MS, Ferraro M, Izaguirre I, Pizarro H. "New findings on the effect of glyphosate on autotrophic and heterotrophic picoplankton structure: A microcosm approach." Aquat Toxicol. 2020 May;222:105463. doi: 10.1016/j.aquatox.2020.105463. Epub 2020 Mar 3. PMID: 32172181.

Sagiv, Sharon & Kogut, Katherine & Harley, Kim & Bradman, Asa & Morga, Norma & Eskenazi, Brenda. (2021). "Gestational exposure to organophosphate pesticides and longitudinally assessed behaviors related to ADHD and executive function." American Journal of Epidemiology. 190. 10.1093/aje/kwab173.

Saini A, Aggarwal NK, Sharma A, Yadav A. Actinomycetes: A Source of Lignocellulolytic Enzymes. Enzyme Res. 2015;2015:279381. doi: 10.1155/2015/279381. Epub 2015 Dec 17. PMID: 26793393; PMCID: PMC4697097.

Salas JH, González MM, Noa M, Pérez NA, Díaz G, Gutiérrez R, Zazueta H, Osuna I. "Organophosphorus pesticide residues in Mexican commercial pasteurized milk." J Agric Food Chem. 2003 Jul 16;51(15):4468-71. doi: 10.1021/jf020942i. PMID: 12848527.

Saleh Alwaneen W. "Effect of Cow Manure Compost on Chemical and Microbiological Soil Properties in Saudi Arabia." Pak J Biol Sci. 2020 Jan;23(7):940-945. doi: 10.3923/pjbs.2020.940.945. PMID: 32700842.

Salkida, Ahmad, "Africa's vanishing Lake Chad", United Nations, Africa Renewal, April 2012, https://www.un.org/africarenewal/magazine/april-2012/africa%E2%80%99s-vanishing-lake-chad

Samsel A, Seneff S. "Glyphosate, pathways to modern diseases III: Manganese, neurological diseases, and associated pathologies." Surg Neurol Int. 2015 Mar 24;6:45. doi: 10.4103/2152-7806.153876. PMID: 25883837; PMCID: PMC4392553.

Sanders, Robert, "Fertilizer use responsible for increase in nitrous oxide in atmosphere" UC Berkeley, 2012, https://news.berkeley.edu/2012/04/02/fertilizer-use-responsible-for-increase-in-nitrous-oxide-in-atmosphere/

Santina Zancan, Renata Trevisan, Maurizio G. Paoletti, "Soil algae composition under different agro-ecosystems in North-Eastern Italy," Agriculture, Ecosystems & Environment, Volume 112, Issue 1, 2006, Pages 1-12, https://doi.org/10.1016/j.agee.2005.06.018.

Santos Sánchez, Norma & Salas-Coronado, Raul & Hernández-Carlos, Beatriz & Villanueva, Claudia. (2019). "Shikimic Acid Pathway in Biosynthesis of Phenolic Compounds." 10.5772/intechopen.83815.

Schafer, Sara, "How Many Farms Are in the U.S.?" Farm Journal, Ag Web, 2023, https://www.agweb.com/news/business/farmland/how-many-farms-are-us

Schimel, Joshua & Bennett, Jennnifer. (2004). "Nitrogen Mineralization: Challenges of a Changing Paradigm." Ecology. 85. 591–602. 10.1890/03-8002.

Scholtz MT, Bidleman TF. "Modelling of the long-term fate of pesticide residues in agricultural soils and their surface exchange with the atmosphere: Part II. Projected long-term fate of pesticide residues." Sci Total Environ. 2007 May 1;377(1):61-80. doi: 10.1016/j.scitotenv.2007.01.084. Epub 2007 Mar 8. PMID: 17346778.

Schreiner, R. P., & Koide, R. T. (1993). Streptomycin reduces plant response to mycorrhizal infection. Soil Biology and Biochemistry, 25(8), 1131–1133. https://doi.org/10.1016/0038-0717(93)90162-5

Schulz D, Nachtigall J, Riedlinger J, Schneider K, Poralla K, Imhoff JF, Beil W, Nicholson G, Fiedler HP, Süssmuth RD. "Piceamycin and its N-acetylcysteine adduct is produced by Streptomyces" sp. GB 4-2. J Antibiot (Tokyo). 2009 Sep;62(9):513-8. doi: 10.1038/ja.2009.64. Epub 2009 Jul 17. PMID: 19609293.

Science Direct, "Acetylcholinesterase", 2023, https://www.sciencedirect.com/topics/agricultural-and-biological-sciences/acetylcholinesterase

Science Direct, "Atmospheric Deposition", 2023, https://www.sciencedirect.com/topics/earth-and-planetary-sciences/atmospheric-deposition

Science Direct, "Mortierella alpina", 2023, https://www.sciencedirect.com/topics/biochemistry-genetics-and-molecular-biology/mortierella-alpina

Science Direct, "Scenedesmus", 2023, https://www.sciencedirect.com/topics/earth-and-planetary-sciences/scenedesmus

Scoones, Ian, Reij, Chris, and Toulmin, Camilla, "Sustaining the Soil Indigenous Soil and Water Conservation in Africa" Routledge. 1996

Scribner, Elisabeth A., Battaglin, William A., Gilliom, Robert J., and Meyer, Michael T., "Concentrations of Glyphosate, Its Degradation Product, Aminomethylphosphonic Acid, and Glufosinate in Ground- and Surface-Water, Rainfall, and Soil Samples Collected in the United States, 2001–06" U.S. Department of the Interior, U.S. Geological Survey, Scientific Investigations Report 2007–5122 https://pubs.usgs.gov/sir/2007/5122/pdf/SIR2007-5122.pdf

Sellars, S. and V. Nunes. "Synthetic Nitrogen Fertilizer in the U.S." farmdoc daily (11):24, Department of Agricultural and Consumer Economics,University of Illinois at Urbana-Champaign, February 17, 2021.

https://farmdocdaily.illinois.edu/2021/02/synthetic-nitrogen-fertilizer-in-the-us.html

Seneff, Stephanie & Swanson, Nancy & Li, Chen. (2015). "Aluminum and Glyphosate Can Synergisti-cally Induce Pineal Gland Pathology:" Connection to Gut Dysbiosis and Neurological Disease. Agricultural Sciences. 06. 42-70. 10.4236/as.2015.61005.

Sengupta, S., Pramanik, A., Ghosh, A. et al. "Antimicrobial activities of actinomycetes isolated from unexplored regions of Sundarbans mangrove ecosystem." BMC Microbiol 15, 170 (2015). https://doi.org/10.1186/s12866-015-0495-4

Shahabifar, Jafar & Panahpoor, Ebrahim & Moshiri, Farhad & Gholami, Ali & Mostashari, Mehrzad. (2019). "The quantity/intensity relation is affected by chemical and organic P fertilization in calcareous soils." Ecotoxicology and Environmental Safety. 172. 144-151. 10.1016/j.ecoenv.2019.01.058.

Shanab, Sanaa & Hafez, Rehab & Fouad, Ahmed. (2018). "A review on algae and plants as potential source of Arachidonic acid." Journal of Advanced Research. 2018. 3-13. 10.1016/j.jare.2018.03.004.

Shaner, D.L. (2004), "Herbicide safety relative to common targets in plants and mammals." Pest. Manag. Sci., 60: 17-24. https://doi.org/10.1002/ps.782

Sharkey, Stephen M. ; Hartig, Anna M. ; Dang, Audrey J. ; Chatterjee, Anamika ; Williams, Brent J. ; Parker, Kimberly M., "Amine Volatilization from Herbicide Salts: Implications for Herbicide Formulations and Atmospheric Chemistry", Environmental Science & Technology, vol. 56, issue 19, pp. 13644-13653, 2022, DOI: 10.1021/acs.est.2c03740

Shi, L., Steenland, K., Li, H. et al. "A national cohort study (2000–2018) of long-term air pollution exposure and incident dementia in older adults in the United States." Nat Commun 12, 6754 (2021). https://doi.org/10.1038/s41467-021-27049-2

Shruthi PJ, Sujatha K, Srilatha CH, et al. "Incidence of different tumors in bovines." Open Access J Sci. 2018;2(4):220-222. DOI: 10.15406/oajs.2018.02.00076

Shwartz, Mark, "Study highlights massive imbalances in global fertilizer use", Phys.org, June 22, 2009, https://phys.org/news/2009-06-highlights-massive-imbalances-global-fertilizer_1.html

Sillito, John, "A Utahn Abroad: Parley P. Christensen's World Tour, 1921-23" Utah Historical Quarterly, Volume 54, Number 4, 1986, https://issuu.com/utah10/docs/uhq_volume54_1986_number4/s/152868

Simopoulos AP. "Dietary omega-3 fatty acid deficiency and high fructose intake in the development of metabolic syndrome, brain metabolic abnormalities, and non-alcoholic fatty liver disease." Nutrients. 2013 Jul 26;5(8):2901-23. doi: 10.3390/nu5082901. PMID: 23896654; PMCID: PMC3775234.

Singh, U., Walvekar, V.A. and Sharma, S. (2020). "Microbiome as Sensitive Markers for Risk Assessment of Pesticides. In Pesticides in Crop Production" (eds P.K. Srivastava, V.P. Singh, A. Singh, D.K. Tripathi, S. Singh, S.M. Prasad and D.K. Chauhan). https://doi.org/10.1002/9781119432241.ch6

Skaper SD, Di Marzo V. "Endocannabinoids in nervous system health and disease: the big picture in a nutshell." Philos Trans R Soc Lond B Biol Sci. 2012 Dec 5;367(1607):3193-200. doi: 10.1098/rstb.2012.0313. PMID: 23108539; PMCID: PMC3481537.

Smith, Jeffery M., "Genetic Roulette" Yes! Books 2007

Smithsonian Education, "Buying Time", 2013, https://smithsonianeducation.org/educators/lesson_plans/revolutionary_money/intro_3.html

Sochocka M, Karska J, Pszczołowska M, Ochnik M, Fułek M, Fułek K, Kurpas D, Chojdak-Łukasiewicz J, Rosner-Tenerowicz A, Leszek J. "Cognitive Decline in Early and Premature Menopause." International Journal of Molecular Sciences. 2023; 24(7):6566. https://doi.org/10.3390/ijms24076566

Socorro, Joanna & Gligorovski, Sasho & Wortham, Henri & Quivet, Etienne. (2015). "Heterogeneous reactions of ozone with commonly used pesticides adsorbed on silica particles."

Atmospheric Environment. 100. 66-73. 10.1016/j.atmosenv.2014.10.044.
Sonne J, Reddy V, Beato MR. "Neuroanatomy, Substantia Nigra." [Updated 2022 Oct 24]. In: StatPearls [Internet]. Treasure Island (FL): StatPearls Publishing; 2023 Jan-. Available from: https://www.ncbi.nlm.nih.gov/books/NBK536995/
Soth, L., (1983) "Henry Wallace and the Farm Crisis of the 1920S and 1930S", The Annals of Iowa 47(2), 195-214. doi: https://doi.org/10.17077/0003-4827.8993
Sparks TC, Nauen R. "IRAC: Mode of action classification and insecticide resistance management." Pestic Biochem Physiol. 2015 Jun;121:122-8. doi: 10.1016/j.pestbp.2014.11.014. Epub 2014 Dec 4. PMID: 26047120.
Spiegel, Bill, "Companion Crops Boost Wheat Stubble Health", Successful Farming, 1/28/2015, https://www.agriculture.com/crops/wheat/production/com pion-crops-boost-wheat-stubble_145-ar47222
Spohn, M.. (2015). "Microbial respiration per unit microbial biomass depends on soil litter carbon-to-nitrogen ratio." Biogeosciences. 12. 817-823. 10.5194/bg-12-817-2015.
Starrett, Henry P., "East African Markets for Hardware and Agricultural Implements", U.S. Government Printing Office, 1917
State of Louisiana, Department of Ag. and Forestry, "Mississippi River siltation issues continue to be problem" June 8, 2011 https://www.ldaf.state.la.us/news/mississippi-river-siltation-issues-continue-to-be-problem/
Stefano Bedini, Elisa Pellegrino, Luciano Avio, Sergio Pellegrini, Paolo Bazzoffi, Emanuele Argese, Manuela Giovannetti, "Changes in soil aggregation and glomalin-related soil protein content as affected by the arbuscular mycorrhizal fungal species Glomus mosseae and Glomus intraradices" Soil Biology and Biochemistry, Volume 41, Issue 7, July 2009, Pages 1491-1496 https://doi.org/10.1016/j.soilbio.2009.04.005
Stern, Caryl M., "A Slow-Motion Crisis: Gen Z's Battle Against Depression, Addiction, Hopelessness", The 74, 2022, https://www.the74million.org/article/a-slow-motion-crisis-gen-zs-battle-against-depression-addiction-hopelessness/

Strilbyska OM, Tsiumpala SA, Kozachyshyn II, Strutynska T, Burdyliuk N, Lushchak VI, Lushchak O. "The effects of low-toxic herbicide Roundup and glyphosate on mitochondria." EXCLI J. 2022 Jan 10;21:183-196. doi: 10.17179/excli2021-4478. PMID: 35221840; PMCID: PMC8859649.

Sylwestrzak Z, Zgrundo A, Pniewski F. "Ecotoxicological Studies on the Effect of Roundup® (Glyphosate Formulation) on Marine Benthic Microalgae." Int J Environ Res Public Health. 2021 Jan 20;18(3):884. doi: 10.3390/ijerph18030884. PMID: 33498564; PMCID: PMC7908156.

Tan, Kim H., "Humic Matter in Soil and the Environment Principles and Controversies, Second Edition" CRC Press, 2014

Tegeder M. "Transporters involved in source to sink partitioning of amino acids and ureides: opportunities for crop improvement." J Exp Bot. 2014 Apr;65(7):1865-78. doi: 10.1093/jxb/eru012. Epub 2014 Jan 31. PMID: 24489071.

Temple, James, "What is geoengineering—and why should you care?" MIT Technology Review, 2019, https://www.technologyreview.com/2019/08/09/615/what-is-geoengineering-and-why-should-you-care-climate-change-harvard/

Teng, Hongfen & Viscarra Rossel, Raphael & Shi, Zhou & Behrens, Thorsten & Chappell, Adrian & Bui, Elisabeth. (2016). "Assimilating satellite imagery and visible–near infrared spectroscopy to model and map soil loss by water erosion in Australia." Environmental Modelling & Software. 77. 156-167. 10.1016/j.envsoft.2015.11.024.

The Annie E. Casey Foundation , "Generation Z and Mental Health", 2023, https://www.aecf.org/blog/generation-z-and-mental-health

The Cattle Site, "Dietary Essential Fatty Acids and Reproduction in Dairy Cows", 17 February 2009, https://www.thecattlesite.com/articles/1615/dietary-essential-fatty-acids-and-reproduction-in-dairy-cows

The Louisiana Coastal Wetlands Planning Protection and Restoration Act Program "The Mississippi River Delta Basin" https://lacoast.gov/new/about/basin_data/mr/

The Recovery Village, "Gender Dysphoria Statistics", 2023, https://www.therecoveryvillage.com/mental-health/gender-dysphoria/gender-dysphoria-statistics/

Thi Quynh Anh, Pham & Gomi, Takashi & MacDonald, Lee & Mizugaki, Shigeru & Khoa, Phung & Furuichi, Taka. (2014). "Linkages among land use, macronutrient levels, and soil erosion in northern Vietnam: A plot-scale study." Geoderma. s 232–234. 352–362. https://doi.org10.1016/j.geoderma.2014.05.011.

Thiermann H, Worek F. "Pro: Oximes should be used routinely in organophosphate poisoning" Br J Clin Pharmacol. 2022 Dec;88(12):5064-5069. doi: 10.1111/bcp.15215. Epub 2022 Feb 7. PMID: 35023196

Tian Y, Feng F, Zhang B, Li M, Wang F, Gu L, Chen A, Li Z, Shan W, Wang X, Chen X, Zhang Z. "Transcriptome analysis reveals metabolic alteration due to consecutive monoculture and abiotic stress stimuli in Rehamannia glutinosa" Libosch. Plant Cell Rep. 2017 Jun;36(6):859-875. doi: 10.1007/s00299-017-2115-2. Epub 2017 Mar 8. PMID: 28275853

Tilev, Şeyda & Kahraman, Abdullah. (2015). Saharan dust transport by Mediterranean cyclones causing mud rain in Istanbul. Weather. 70. 10.1002/wea.2472.

Tinta, T., Klun, K. and Herndl, G.J. (2021), "The importance of jellyfish–microbe interactions for biogeochemical cycles in the ocean." Limnol Oceanogr, 66: 2011-2032. https://doi.org/10.1002/lno.11741

Tongo I, Ezemonye L. "Human health risks associated with residual pesticide levels in edible tissues of slaughtered cattle in Benin City, Southern Nigeria." Toxicol Rep. 2015 Aug 3;2:1117-1135. doi: 10.1016/j.toxrep.2015.07.008. PMID: 28962453; PMCID: PMC5598159.

Tuchscherer, Rebekah, "Study finds residue of pesticides, antibiotics and growth hormone in non-organic milk", USA Today, 2019, https://www.usatoday.com/story/money/2019/06/26/study-finds-organic-milk-cleaner-than-conventional-dairy/1482508001/

Tyson, Peter, "The Contrail Effect", NOVA, 2006, https://www.pbs.org/wgbh/nova/sun/contrail.html

Tzfira, T & Hohn, B. & Gelvin, Stanton. (2017). Transfer of Genetic Information From Agrobacterium to Plants. 10.1016/B978-0-12-809633-8.07278-2.

U.S. Government Accountability Office, "Farm Bill: Reducing Crop Insurance Costs Could Fund Other Priorities", 2023, https://www.gao.gov/products/gao-23-106228

University of California, Division of Agricultural and Natural Resources, "Long Chain Fatty Acid Inhibitors", 2023, https://herbicidesymptoms.ipm.ucanr.edu/MOA/Long_Chain_Fatty_Acid_Inhibitors/

University of California-Davis, "Why insect pests love monocultures, and how plant diversity could change that", Science Daily, Oct. 12th, 2016

University of Exeter, "Charting the Global Invasion of Crop Pests", Science Daily, Aug. 27th, 2014

University of Illinois at Urbana-Champaign, "Nematode resistance in soybeans beneficial even at low rates of infestation", February 23, 2017, https://phys.org/news/2017-02-nematode-resistance-soybeans-beneficial-infestation.html

University of Minnesota Extension, "Cell membrane disruption herbicides", 2018, https://extension.umn.edu/herbicide-mode-action-and-sugarbeet-injury-symptoms/cell-membrane-disruption-herbicides#bipyridyliums-1665865

University of Minnesota Extension, "Fertilizer urea", 2021, https://extension.umn.edu/nitrogen/fertilizer-urea#losing-urea-due-to-soil-temperature-and-ph-755164

University of Minnesota Extension, "Growth regulator herbicides", 2018, https://extension.umn.edu/herbicide-mode-action-and-sugarbeet-injury-symptoms/growth-regulator-herbicides#auxin-transport-inhibitor-1657863

USDA, "Food Imports" 2016 https://www.ers.usda.gov/data-products/us-food-imports/

USDA, ERS, "Farm Bill Spending" 2023, https://www.ers.usda.gov/topics/farm-economy/farm-commodity-policy/farm-bill-spending/

USDA, ERS, "History of Agricultural Price-Support and Adjustment Programs 1933-84", Bulletin Number 485, 1984, https://www.ers.usda.gov/webdocs/publications/41988/50849_aib485.pdf

USDA, ERS, "The Role of Changing Vertical Coordination in the Broiler and Pork Industries", 1999, https://www.ers.usda.gov/webdocs/publications/40999/17957_aer777b_1_.pdf?v=0

USDA, ERS, "U.S. fruit and vegetable import value outpaces volume growth" July 13, 2022, https://www.ers.usda.gov/data-products/chart-gallery/gallery/chart-detail/?chartId=104212

USDA, FAS, "Exporter Guide", Report Number: LY2020-0002, March 06,2020, https://apps.fas.usda.gov/newgainapi/api/Report/DownloadReportByFileName?fileName=Exporter%20Guide_Rabat_Libya_03-06-2020

USDA, FAS, "Grain: World Markets and Trade", May 2023, https://apps.fas.usda.gov/psdonline/circulars/grain.pdf

USDA, FAS, ERS, AMS, FSA, Office of the Chief Economist, "World Agricultural Supply and Demand Estimates", November 9, 2022, https://downloads.usda.library.cornell.edu/usda-esmis/files/3t945q76s/3j334c74p/fx71bx60x/wasde1122.pdf

USDA, NASS, "2019 AGRICULTURAL CHEMICAL USE SURVEY", No. 2020-5, 2020, https://www.nass.usda.gov/Surveys/Guide_to_NASS_Surveys/Chemical_Use/2019_Field_Crops/chem-highlights-wheat-2019.pdf

USGS, "Nitrogen Statistics and Information", 2011, https://www.usgs.gov/centers/national-minerals-information-center/nitrogen-statistics-and-information

Verheijen, F.G.A., Jones, R.G.A., Rickson, R.J., Smith, C.J., "Tolerable versus actual soil erosion rates in Europe," Earth-Science Reviews, Volume 94, Issues 1–4, 2009, Pages 23-38, https://doi.org/10.1016/j.earscirev.2009.02.003.

Verzeñassi, Damián, Vallini, Alejandro, Fernández, Facundo, Ferrazini, Lisandro, Lasagna, Marianela, Sosa, Anahí J., Hough, Guillermo E., "Cancer incidence and death rates in Argentine rural towns surrounded by pesticide-treated agricultural land," Clinical Epidemiology and Global Health, Vol. 20, 2023, https://doi.org/10.1016/j.cegh.2023.101239.

Waits A, Chang CH, Yu CJ, Du JC, Chiou HC, Hou JW, Yang W, Chen HC, Chen YS, Hwang B, Chen ML. "Exposome of attention deficit hyperactivity disorder in Taiwanese children: exploring risks of endocrine-disrupting chemicals." J Expo Sci Environ Epidemiol. 2022 Jan;32(1):169-176. doi: 10.1038/s41370-021-00370-0. Epub 2021 Jul 15. PMID: 34267309.

Waksman, Selman A. , "Soil Microbiology" John Wiley and Sons, 1952

Walorczyk, Stanislaw & Drożdżyński, Dariusz. (2012). "Improvement and extension to new analytes of a multi-residue method for the determination of pesticides in cereals and dry animal feed using gas chromatography-tandem quadrupole mass spectrometry revisited." Journal of chromatography. A. 1251. 219-31. 10.1016/j.chroma.2012.06.055.

Walsh, D.F., Berger, B.L., & Bean, J.R. (1977). "Mercury, arsenic, lead, cadmium, and selenium residues in fish, 1971-73" National Pesticide Monitoring Program. Pesticides monitoring journal, 11 1, 5-34 .

Wan, Xiaohua & Huang, Zhiqun & He, Zongming & Yu, Zaipeng & Wang, Minhuang & Davis, Murray & Yang, Yusheng. (2014). "Soil C:N ratio is the major determinant of soil microbial community structure in subtropical coniferous and broadleaf forest plantations." Plant and Soil. 387. 10.1007/s11104-014-2277-4.

Wang, J. & Zou, Baoping & Liu, Yan & Tang, Yiqun & Zhang, Xinbao & Yang, Ping. (2014). "Erosion-creep-collapse mechanism of underground soil loss for the karst rocky desertification in Chenqi village, Puding county, Guizhou, China." Environmental Earth Sciences. 72. 2751-2764. 10.1007/s12665-014-3182-0.

Wang, Ning & Besser, John & Buckler, Denny & Honegger, Joy & Ingersoll, Chris & Johnson, B.T. & Kurtzweil, Mitchell & Macgregor, Jon & McKee, Michael. (2005). "Influence of sediment on the fate and toxicity of a polyethoxylated tallowamine surfactant system (MON 0818) in aquatic microcosms." Chemosphere. 59. 545-551. 10.1016/j.chemosphere.2004.12.009.

Wang, S., Nan, J., Shi, C. et al. "Atmospheric ammonia and its impacts on regional air quality over the megacity of Shanghai, China." Sci Rep 5, 15842 (2015). https://doi.org/10.1038/srep15842

Wani AL, Bhat SA, Ara A. "Omega-3 fatty acids and the treatment of depression: a review of scientific evidence." Integr Med Res. 2015 Sep;4(3):132-141. doi: 10.1016/j.imr.2015.07.003. Epub 2015 Jul 15. PMID: 28664119; PMCID: PMC5481805.

Water Resources Mission Area, "Glyphosate: Widely Used, Widely Detected", USGS, 2020, https://www.usgs.gov/news/herbicide-glyphosate-prevalent-us-streams-and-rivers

Weaver, L. M., & Herrmann, K. M. (1997). "Dynamics of the shikimate pathway in plants." Trends in Plant Science, 2(9), 346–351. https://doi.org/10.1016/S1360-1385(97)84622-5

Welsh, J., Braun, H., Brown, N., Um, C., Ehret, K., Figueroa, J., & Boyd Barr, D. (2019). Production-related contaminants (pesticides, antibiotics and hormones) in organic and conventionally produced milk samples sold in the USA. Public Health Nutrition, 22(16), 2972-2980. doi:10.1017/S136898001900106X

Whiteman, Lily, "Jellyfish Gone Wild", National Science Foundation, 2023, https://www.nsf.gov/news/special_reports/jellyfish/NSF_JellyfishGoneWild.pdf

Williams, A., Ridgway, H.J. & Norton, D.A. "Different arbuscular mycorrhizae and competition with an exotic grass affect the growth of Podocarpus cunninghamii Colenso cuttings." New Forests 44, 183–195 (2013). https://doi.org/10.1007/s11056-012-9309-9

Woodrow JE, Gibson KA, Seiber JN. "Pesticides and Related Toxicants in the Atmosphere." Rev Environ Contam Toxicol. 2019;247:147-196. doi: 10.1007/398_2018_19. PMID: 30535549.

World Meteorological Organization (WMO), "Atmospheric Deposition", 2022, https://public.wmo.int/en/our-mandate/focus-areas/environment/atmospheric-deposition

Worster, Donald, "Dust Bowl", Texas State Historical Association, 2017 https://www.tshaonline.org/handbook/entries/dust-bowl

Wright, Sara and Nichols, Kristine, "Glomalin: Hiding Place for a Third of the World's Stored Soil Carbon", USDA ARS AgMag., 2002

Wu, D., Jiang, L., Li, W., Wu, Q., & Lv, G. (2023). "Drivers of rhizosphere microbial differences in desert genus Haloxylon." Land Degradation & Development, 1– 12. https://doi.org/10.1002/ldr.4699

Xiang, Xingjia & Gibbons, Sean & He, Jin-Sheng & Wang, Chao & He, Dan & Li, Qian & Ni, Yingying & Chu, Haiyan. (2016). "Rapid response of arbuscular mycorrhizal fungal communities to short-term fertilization in an alpine grassland on the Qinghai-Tibet Plateau." PeerJ. 4. e2226. 10.7717/peerj.2226.

Yang H, Ye W, Ma J, Zeng D, Rong Z, Xu M, Wang Y, Zheng X. "Endophytic fungal communities associated with field-grown soybean roots and seeds in the Huang-Huai region of China." PeerJ. 2018 Apr 30;6:e4713. doi: 10.7717/peerj.4713. PMID: 29736345; PMCID: PMC5933319.

Yao Pan and others, Impact of long-term N, P, K, and NPK fertilization on the composition and potential functions of the bacterial community in grassland soil, FEMS Microbiology Ecology, Volume 90, Issue 1, October 2014, Pages 195–205, https://doi.org/10.1111/1574-6941.12384

Yirka, Bob, "Study shows Lake Mega-Chad dried up far more quickly than thought" , Phys.org, June 30, 2015, https://phys.org/news/2015-06-lake-mega-chad-dried-quickly-thought.html

Young SN. "Acute tryptophan depletion in humans: a review of theoretical, practical and ethical aspects." J Psychiatry Neurosci. 2013 Sep;38(5):294-305. doi: 10.1503/jpn.120209. PMID: 23428157; PMCID: PMC3756112.

Zancan, Santina & Trevisan, Renata & Paoletti, Maurizio. (2006). Soil algae composition under different agro-ecosystems in North-Eastern Italy. Agriculture, Ecosystems & Environment. 1-12. 10.1016/j.agee.2005.06.018.

Zhang S, Xu J, Kuang X, Li S, Li X, Chen D, Zhao X, Feng X. "Biological impacts of glyphosate on morphology, embryo biomechanics and larval behavior in zebrafish (Danio rerio)." Chemosphere. 2017 Aug;181:270-280. doi: 10.1016/j.chemosphere.2017.04.094. Epub 2017 Apr 21. PMID: 28448908.

Zhang, Changpeng & Liu, Xingang & Dong, Fengshou & Xu, Jun & Zheng, Yong-Quan & Li, Jing. (2010). "Soil microbial communities response to herbicide 2,4-dichlorophenoxyacetic acid butyl ester." European Journal of Soil Biology. 46. 175-180. 10.1016/j.ejsobi.2009.12.005.

Zhang, H., Cui, Q. & Song, X. "Research advances on arachidonic acid production by fermentation and genetic modification of Mortierella alpina." World J Microbiol Biotechnol 37, 4 (2021). https://doi.org/10.1007/s11274-020-02984-2

Zhang, Weidong & Wang, Jiachao & Song, Jianshi & Feng, Yanru & Zhang, Shujuan & Wang, Na & Liu, Shufeng & Song, Zhixue & Lian, Kaoqi & Kang, Weijun. (2021). "Effects of low-concentration glyphosate and aminomethyl phosphonic acid on zebrafish embryo development." Ecotoxicology and Environmental Safety. 226. 112854. 10.1016/j.ecoenv.2021.112854.

Zhiyong, Li, "A policy review on watershed protection and poverty alleviation by the Grain for Green Programme in China" China Academy of Forestry https://www.fao.org/3/ae537e/ae537e0j.htm

Zhong W, Xu D, Chai Z, Mao X. "2001 survey of organochlorine pesticides in retail milk from Beijing, P R China." Food Addit Contam. 2003 Mar;20(3):254-8. doi: 10.1080/0265203021000055405. PMID: 12623650.

Zolezzi, J. M., Bastías-Candia, S., & Inestrosa, N. C., "Molecular Basis of Neurodegeneration: Lessons from Alzheimer's and Parkinson's Diseases." (2019). IntechOpen. doi: 10.5772/intechopen.81270

www.ingramcontent.com/pod-product-compliance
Lightning Source LLC
Chambersburg PA
CBHW031617210526
45464CB00004B/1612